Air conditioning and energy conservation

ORGANISING COMMITTEE

Chairman
A. F. C. Sherratt, BSc, PhD, CEng, FIMechE, FCIBS, FInstR

Colin Izzard, FCIBS, MConsE (representing the Chartered Institution of Building Services)

W. Peter Jones, MSc, CEng, FIHVE, FInstF, MASHRAE (representing the Chartered Institution of Building Services)

Edward James Perry, CEng, MIMechE, FInstR (representing the Institute of Refrigeration)

Deryck Thornley, FCIBS, MASHRAE (representing the Chartered Institution of Building Services)

R. B. Watson, FRICS (representing the Royal Institution of Chartered Surveyors)

Air conditioning and energy conservation

Edited by A. F. C. Sherratt

The Architectural Press: London

This book is the proceedings of the conference organised in April, 1978 at the University of Nottingham by the Construction Industry Conference Centre Ltd in conjunction with the Chartered Institution of Building Services, the Royal Institution of Chartered Surveyors and the Institute of Refrigeration

First published in book form in 1980 by The Architecural Press Ltd: London

British Library Cataloguing in Publication Data

Conference on Air Conditioning and Energy
  Conservation, *University of Nottingham, 1978*
  Air conditioning and energy conservation.
  1. Air conditioning – Equipment and supplies –
  Design and construction – Congresses
  2. Buildings – Energy conservation – Congresses
  I. Title II. Sherratt, Alan Frederick Cave
  III. Construction Industry Conference Centre
  697.9′312   TH7687.5
  ISBN 0 85139 043 9

Library of Congress Cataloging in Publication Data

Main entry under title:

Air conditioning and energy conservation.

  1. Air conditioning – Congresses. 2. Buildings –
  Energy conservation – Congresses. I. Construction
  Industry Conference Centre.
  TH7687.A1A42   697.9′3   79–18942
  ISBN 0–89397–071–9

Printed and bound in Great Britain by W & J Mackay Limited, Chatham, Kent

Contents

Acknowledgements
The editor and Technical Organising Committee wish to thank everyone who has participated in the arrangements and operation of the conference and in the production of these proceedings. A special word of thanks is given to Mrs Diana Bell and her colleagues at the Construction Industry Conference Centre for their efficiency and helpfulness.

Preface

Air conditioning is the best means known to man of providing fine control of the environment inside buildings. The inherent advantages of an air-conditioned building have resulted in greater demand for air-conditioning from building owners and users, and over the past thirty years the number of installations has increased considerably. This has happened all over the world, both in tropical and temperate climates, although the reasons for the developments are different in the two zones.

Energy use was not a criterion of great importance, energy was cheap and even if optimisation exercises had been carried out, the answers would have been very different from those obtained today. The man-made energy crisis in 1973 changed all that. The cost of energy suddenly jumped, the price in real terms is still rising and by the turn of the century it will probably be double what it is today. As important, though, was the impact on the general awareness of both professionals and laymen that fossil fuel reserves are finite and are being used up at a tremendous rate. Almost overnight energy conservation became a vital issue and furthermore a financial proposition.

This book is about designing air-conditioned buildings in which the use of energy is minimised. Against an energy conservation background, the criteria for air conditioning are examined, the design of air-conditioning systems re-assessed and the implications on building design and on costs investigated. The chapters, contributed by leading experts, provide a great deal of valuable information, including several important case studies of buildings designed to optimise the use of energy and other resources.

The Conference on Air Conditioning and Energy Conservation, of which this book is the proceedings, gave an opportunity for an in-depth appraisal of air conditioning in a new environment. It is hoped that the data and experiences documented both in the papers and in the discussion will be of direct use to designers of air-conditioning systems in both new and existing buildings.

A. F. C. Sherratt
London

It is with sincere regret that I record the death of Lionel Ginsler in August 1978 from injuries sustained in a car accident: a great loss to his many friends and colleagues and to the air-conditioning industry.

A.F.C.S.

List of contributors

David Allford, BA(Arch), FRIBA, FRAIA, FFB, Partner, Yorke Rosenberg Mardall, London
E. J. Anthony, BSc Eng, CEng, MIMechE, MIEE, Director, W. S. Atkins & Partners, Epsom, Surrey
Dr John Cunningham, MP, Parliamentary Under-Secretary of State at the Department of Energy
P. H. Day, C Eng, MIMechE, MCIBS, MIPlantE, Deputy Chief Engineer, City of London Real Property Co Ltd
John P. Eberhard, President, Architects' Institute of America Research Corporation, Washington DC
L. S. Ginsler, P Eng, BASc, MEIC, President, Rybka Smith & Ginsler Ltd, Consulting Engineers, Toronto, Canada
J. P. G. Goldfinger, MA (Cantab), FCIBS, MASHRAE, MConsE, Partner, Dale and Goldfinger
V. A. Hammond, FCIBS, FI Hosp E, Partner, Steensen Varming Mulcahy & Partners, Hemel Hempstead, Herts
M. J. Holmes, BSc, ACGI, DIC, Ove Arup Partnership, formerly head of the Systems Performance Unit, Building Services Research and Information Association, Bracknell
A. Leslie Longworth, MSc (Tech), PhD, CEng, FIMechE, MICE, FCIBS, M Inst F, President, Chartered Institution of Building Services, Senior Lecturer, University of Manchester, Institute of Science and Technology
E. C. Lovelock, CEng, MIMechE, Manager, Estates Services Division, Shell UK Ltd
N. Nossan, DEng, MAssEng (Italy), General Manager, Aster International, Milan, Italy
G. Pancaldi, DEng, MAssEng (Italy), Technical Manager, Aster International, Milan, Italy
Jack Peach, BSc, CEng, FIMechE, FCIBS, MIEE, Technical Secretary, the Chartered Institution of Building Services
Edward James Perry, CEng, MIMechE, FInstR, Chief Engineer, W. S. Atkins & Partners, Epsom, Surrey
Sir Alan Pullinger, CBE, Chairman, Haden Carrier Ltd, London
A. F. C. Sherratt, BSc, PhD, CEng, FIMechE, FCIBS, FInstR, Assistant Director & Dean of the Faculty of Architecture and Surveying, Thames Polytechnic, London
P. R. Shiner, BSc, MCIBS, Director, Dale & Ewbank, London
T. Smith, CEng, FIMechE, FInstF, FCIBS, MConsE, Partner, Steensen Varming Mulcahy & Partners, Hemel Hempstead, Herts
P. C. Venning, FRICS, Chairman, Engineering Services Committee, Royal Institution of Chartered Surveyors, Partner, Davis Belfield & Everest, London
R. B. Watson, FRICS, MA Cost E, Associate (Engineering Services), Davis Belfield & Everest, London
J. J. Wilson, MBE, C Eng, FIMarE, FInstR, President, the Institute of Refrigeration, Senior Principal Surveyor, Refrigeration Department, Lloyd's Register of Shipping

Introduction

JOHN CUNNINGHAM

The UK is poised on the edge of energy self-sufficiency. At the same time there is growing awareness both nationally and internationally that world oil supplies will become increasingly scarce and expensive before too long. The London Summit Declaration of May 1977 included a commitment to energy conservation from seven leaders of the industrialised world. Our own Green Paper on Energy Policy, the OECD report *World Energy Outlook*, the Workshop on Alternative Strategies, even the Central Intelligence Agency, have all pointed, in the past year, to the looming problem of world energy supply.

In the UK, the Government has strengthened its commitment to energy conservation as a central element in long term energy policy. The most recent major intitiative – a ten year programme aimed primarily at the public sector and designed to achieve savings equivalent to ten million tonnes of oil by the end of a decade – was only announced in December 1977 and additional measures were set in motion as outlined in the March 1978 Budget: yet further policy additions to what is a continually evolving approach to energy conservation on a wide front.

A British Standards Code of Practice for mechanical ventilation and air conditioning in buildings is in the course of preparation and the chapters and discussion in this book may well make a valuable contribution to the shape of the Code. This will in its turn set the tone for thinking on mechanical ventilation and air conditioning for some time to come. The draft Code of Practice says 'Modern Man desires that his environment should be controlled so that he can be comfortable with the least change of clothing and with the least effort'. In the days before the oil crisis – when adequate secure supplies of cheap oil from the Middle East were taken for granted – we could afford to aim for that level of comfort, that standard of living, for large sections of the community. The first chapter – Low Energy Air Conditioning – highlights the dilemma we now face in the energy world and which the building services industry itself is facing.

Is there a conflict now between what we would like to achieve, in terms of increased comfort and better control of the environment inside buildings, and what we can afford to achieve? Can we achieve the 'comfort in a controlled environment with least effort' of that draft Code, at a price the consumer wishes to pay? This then is the challenge the industry must meet. A fundamental barrier to meeting that challenge is lack of awareness.

1

Although the realisation is slowly growing that world fossil fuel reserves are not only finite but will – certainly in the case of oil – rapidly come under greater pressure as demand continues to increase, this global thought has not yet made sufficient impact at any level. The man in the street either does not believe what he hears experts say, or he does not recognise what it means to him. He often prefers to bury his head in the sand. Neither have many influential people in industry yet grasped the clear consequences of our present patterns of energy use. They do not appreciate what the energy outlook means in their own business operations. The implications of energy scarcity – or of the measures necessary to avert scarcity if we can – have not gone home.

What then is that energy outlook which has such powerful implications for us all? Of course, the immediate future is bright. All the oil we need, plus a probable export surplus, for at least a decade; large quantities of northern basin North Sea gas coming ashore, possibly until the end of the century and beyond; 300 years of technically recoverable coal; 20 years' British experience of safe and viable nuclear generation of electricity.

However, UK oil production is likely to reach a plateau by the mid 80s; offshore gas will probably last a little longer but will be much more difficult to extract from the deeper and more hostile waters of the Northern North Sea – and the price to the consumer will have to reflect those difficulties. We have 300 years supply of coal, but we have to be able to get that coal, at a reasonable price, in the quantities we need, at the time we need it. When our indigenous oil production falls below UK requirements we shall be back in world oil markets competing with America, and the rest of the world for scarcer and more expensive supplies of oil. All these facts point to greater expense in all forms of energy production.

The potential of alternative renewable sources of energy – from the wind, the waves, the sun, tides and hot rocks – is of course being investigated by means of a ten million pounds government research programme. More resources are available to be committed as soon as the potential is proved. There is, however, no immediate solution to the scarcity of traditional forms of energy in these benign sources. The forms of energy sometimes described as 'soft' offer anything but a soft option. Who can say as yet whether they will be acceptable in environmental terms? Who can predict with certainty the environmental effects of the Severn Barrage, or concrete wave power devices off the coasts of Western Scotland? On present estimates, alternative energy will not be cheap. Nuclear fusion – with the promise of infinite safe supplies of energy – is a prospect for the middle of the next century. Our problem will be with us well before then.

There are no easy ways forward in energy supply or energy production. There is, if not an easy option, a more profitable one, in the field of energy demand or energy conservation. It is through the wise and more efficient use of the energy we already produce at huge cost that we can meet the challenge of rising energy prices and dwindling energy resources. It is to energy users we must look for an impact on the energy supply/demand equation.

Energy conservation is by its nature a disaggregated problem. The size of the problem – and its solution – is the result of many factors involving millions of people. Energy use and energy demand is the sum of a million and one decisions, in thousands of different situations. One way of making an impact on demand is to cut out the chance element in those million and one decisions. If we can ensure that any decision to use energy does not result in a decision – however indirectly – to waste energy, we shall have gone a long way down the right road.

It is not only to energy users we must look for energy savings. Manufacturers of energy using equipment – whether buildings, motor cars, boilers, furnaces or air-conditioning units – must provide products which use as little energy as possible to do the job in hand.

Builders and manufacturers must recognise energy efficiency as a new selling point. If they do not recognise it, customers will soon be demanding it as energy prices are expected to double at least in real terms in the next 20 years. Industry must plan now for the consequences of those price rises. To stay in business, industry will have to offer customers efficient, low energy use products.

Most countries are less well placed than ourselves in terms of indigenous energy supplies. Almost all of our industrial competitors are still having to import the energy they need. The pressure of energy efficiency is already on – both in manufacturing processes and in the products they make. Energy labelling on cars and electrical products is already established in France. Britain has followed suit on car labelling. Germany is about to do so on household appliances. Energy efficient products are already on sale – and they may not be British. The message is clear, and cannot be ignored. If British manufacturers are to be successful both at home and abroad, products must be designed for and marketed on economy in energy use.

There is a further consideration for the building service industry, New buildings will inevitably be designed for better fuel use than in the past. They will be better insulated, and better draught-proofed. Regulations for non-domestic buildings, for example, have already been introduced. The heating demand for these better insulated buildings will be lower. Mechanical ventilation for adequate and controlled air changes may well prove to be important as people become aware of the energy costs of open windows and draughty doors. It will be important for manufacturers to develop low energy use air-conditioning units, and there will be a need for smaller, more efficient heating systems. The industry will have to be able to adapt to changing times in the energy world.

DISCUSSION

P. Aris (Marks & Spencer Ltd) The nationalised fuel industries seem to have a policy to sell fuel/energy as quickly as possible without any restraint.

Will the Government take steps to conserve that fuel/energy, in particular gas and oil, without increasing the price?

Dr J. Cunningham (Parliamentary Under-Secretary of State for Energy, Department of Energy)   All the evidence suggests that in a free market economy price has an important role to play in determining the demand for any commodity, and it is clearly the case that movement in energy prices does draw the attention of people to how much energy they are using. In about 1974–5, Eric Varley, then Secretary of State for Energy, invited industry simply to publish in annual reports a statement of energy use. Hardly anyone has done this. People do write long letters to me about the cost of energy. In reply I ask what they are doing to save energy and some think the question either presumptuous, cheek or a typical response from a Minister.

People must fundamentally reappraise the way they use energy, because as Government we do not see any prospect of energy being cheap. North Sea oil is priced at world market prices for oil. It is very expensive to develop, much more expensive than Middle East oil – there is no reason why it should be cheap. Northern Basin North Sea gas must reflect the cost of winning it. I can at the moment see no argument against the economic pricing of energy and that pricing policy will play a role in people's thinking about how they use energy. At the beginning of the present administration the nationalised fuel industries had an accumulated deficit of about 800 million pounds and energy use was being subsidised at a time of energy scarcity. The increase in energy prices has been caused partly by the ending of subsidies paid to the nationalised industries for price restraint, partly by the fivefold increase in oil prices imposed by the OPEC cartel, and partly through the effects of general inflation. Nevertheless, to domestic consumers gas is cheaper now in real terms than at any time in the last eight years. For industry gas is marketed in competition with light fuel oil, and in some respects electricity and coal too, and the Government sees no reason to interfere with operation of the market.

Dr D. Fitzgerald (University of Leeds)   I would like to develop the question of the gas price further. Dr Cunningham said that we cannot demand that fuel prices be other than economic. I would like to know when the price of gas is going to be economic. At the moment it is way below the world market energy price which may be convenient for many of us who pay gas bills, but it is not a good thing for my children. When will the gas price be raised to a proper economic level?

Dr J. Cunningham   It does not make sense to have a fall in real terms in the price of what is a declining commodity although if it is suggested that there should be some kind of thermal parity pricing in world energy terms I do not agree. The ending of subsidies of course did result in increased gas prices. Even if British Gas wanted to raise its prices now they might not be able to because of the Price Commission, and the operation of the law as it stands. Such dilemmas and conflicts cannot be resolved in the next couple of years, but it is a fact that energy prices will have to increase in real terms over a period of time. Although we have a breathing space at present that is no excuse for complacency, but neither is there any benefit in punishing industry or domestic consumers unnecessarily. We already have 1½ million unemployed, a contributing factor towards this being the movement in energy prices.

R. C. Legg (Polytechnic of the South Bank) The cost of energy is far too low at the moment to make much impact on many users. In an air-conditioned office building the cost of the energy for all services (lighting, heating and cooling) amounts to about £1 per week per employee which is perhaps less than ½% of the total cost of employment.

Dr J. Cunningham I cannot comment on the present cost of office heating, suffice it to say that because of movement in prices of fuels it will be greater now than at any time previously. Space and water heating are the major users of energy, and if managers continue to use them as in the past there is something wrong with their management approach.

Sir Alan Pullinger (Haden Carrier Ltd) I like the way Dr Cunningham has talked about pricing in a non-political way. This is a non-political matter. It is no good us complaining that the Government must do this or that on pricing. It is a real dilemma for all of us. Dr Fitzgerald's point is a very real one. The fact is that when one has to decide, particularly in respect of the existing stock of buildings, to make a change with a view to economising in the use of fuel – whether as a private individual or as an enterprise – the decision is made for economic, not moral reasons. The trouble is that there is a glut of oil at the moment, worldwide, and prices are relatively low. In the interest of posterity it would be sensible if, as a nation, we did raise the price of fuel now to make it economical to up-grade installations now, in order to conserve for the future. But we are a competitive nation, and, as Dr Cunningham said, we have a lot of unemployment and one of the best ways to cure that is to export.

Energy represents a large part of the cost of production of many things so if we were to try to encourage conservation by raising fuel prices now we should be in trouble with industrial costs, which would make us internationally uncompetitive. The dilemma is a real one, and the decision will depend on international attitudes to pricing.

Dr J. Cunningham I absolutely agree with Sir Alan, there is a dilemma. The Government in its total approach to energy conservation has deliberately not taken the hard line. It has not sought to control by law, energy use. We have tried to advance by persuasion, by information, technical advice, by financial incentives, to get people to act in their own interest as well as clearly in the national interest. Even though there is something in the argument that we should be increasing energy prices even more, we believe that even now, given the present level of energy prices, investments in energy conservation are one of the best investments any organisation can make. They have a very rapid pay-back, often months, very frequently under a couple of years and there is almost no risk involved in the investment. The lead for more efficient use of resources, in this case energy, must come from the people at the top, without their commitment it is likely that no initiative lower down will succeed.

Sir Alan Pullinger   On the subject of regulations I would like to make a plea for any new regulations to be in a functional rather than a prescriptive form. Technology is moving so fast that if attempts are made to be prescriptive, eg to define in detail automatic controls etc, just about the time all the stages are passed through Parliament, the technology has moved on and the regulation becomes out of date.

Dr J. Cunningham   I am sure that not only this industry but all industry in general will have an interest in the effect of new regulations. Opportunity will be given for written comments on new regulations before any order is made in Parliament.

Mr P. J. Thornton (Rothamsted Experimental Station)   I believe there is a fundamental difficulty in the way Government money is allocated which does not promote energy conservation. I refer to the division between capital expenditure and revenue expenditure with no possibility of transfer between the two. To save revenue, ie energy, it is necessary to obtain the capital to buy the necessary equipment. By altering the way in which its own finance is appropriated Government could take a lead in energy conservation.

Dr J. Cunningham   This is something Government has been trying to do. It is clearly important that capital should be available to invest to enable revenue expenses to be saved. The Property Services Agency, which is responsible for the Government's estate, have by this means made savings running into millions of pounds over the last few years. In December 1977 £320 million were allocated for public sector capital expenditure for a four year energy conservation programme. In industry and commerce a number of other inducements have been tried, the 100% tax allowance for insulating buildings is just one example. My problem as the Minister responsible, is not a shortage of capital, but at least in the public sector, to get that money spent. Government wants the available money to be spent not only to save energy but to galvanise industry and create jobs.

R. Cullen (Architects Design Group)   My practice has been involved in the design of four power stations. Although I believe that 'small is beautiful' I also believe 'big is beautiful', and the construction industry as a whole is short of large projects. Electricity is a great fuel, with cheap electricity we could do all kinds of things very efficiently, for example operate heat pumps and air-conditioning systems. Government should demonstrate its ability to lead and invest some of the money that Dr Cunningham told us is available, to utilise the waste heat from power stations. I know that there are many problems and district heating may not be the answer, but there must be some way of using that waste heat. Holland, for example, uses natural gas to heat greenhouses. Why don't we heat our greenhouses with waste heat and develop agricultural production in a big way, fish farming etc?

Dr J. Cunningham The problem is a real one. Historically large generating stations have been produced with thermal efficiencies in the range of 30–35% and a great deal of waste heat. If industrial processes are taken into account there is much more waste heat. We have established the 'demonstration project scheme', aimed primarily at waste heat recovery in industry, currently £21½ million has been allocated over the next four years to encourage industry to recover and re-cycle waste heat. We are inviting ideas which will be jointly financed, on the understanding that the technology or the results of the demonstration are made available on an industry-wide basis. Using waste heat from generating stations was covered in Energy Paper No 20*, and a more comprehensive report, on combined heat and power and district heating will be received in 1979†.

Work is already underway. The electricity supply industry is engaged in a number of horticultural projects including fish farming and it has recently announced a combined heat and power scheme in Hereford based on a new generating station. The historical problem is not easy to resolve. We have many very large power stations, often close to urban areas. If we were to 'make sure' waste heat was used it would imply not only installing district heating schemes but compulsory take-up, without which the return on investment would not be adequate. In a democracy it is not easy to say to a city all your heating systems are going to be replaced by compulsory district heating. We must make sure, in so far as it is possible, to utilise some of this waste heat; and we must make the most economic use of it in future situations.

*Energy Paper No 20, Department of Energy, HMSO, 1977.
†Energy Paper No 35, Combined Heat and Electrical Power Generated in the UK, Department of Energy, HMSO, 1979.

1 Low energy air conditioning: a challenge for the industry

SIR ALAN PULLINGER

USE OF AIR CONDITIONING

The decision to air condition buildings is usually made to avoid exposure of human beings to excessively high temperatures, both to improve comfort and productivity (eg airports, offices and laboratories), or because people will not patronise establishments if they are uncomfortable (eg theatres, concert halls, hotels and shops), or for therapeutic reasons (operating theatres and other parts of hospitals). Some industrial buildings are air conditioned because the product or plant requires it (eg fine printing, pharmaceutical products, electronic manufacturing). Computer installations also fall into this category.

In the UK, the case for comfort air conditioning scarcely ever rests on modification of the external climate. The number of days when external temperatures exceed the normally accepted level are very few, though the effect of sun on buildings with a high proportion of glazing can be more significant and occur more frequently. Apart from that, the principal reason for installing air conditioning is to counteract internal heat gains, especially those arising from lighting. In places of public assembly humidity control can also be important.

Closed or windowless buildings usually need air conditioning because natural or untreated mechanical ventilation is impractical. It is this factor (ie being closed) more than the building use which is likely to determine whether or not to install air conditioning. Buildings may be closed because of their location, eg in city centres, which require exclusion of external noise, and sometimes dirt too, or because of decisions on shape – cubical buildings generally consume less energy, tall ones introduce exposures to wind. It is clear that there is normally no case for air conditioning in the home.

Although a few people with neurotic tendencies object to air conditioning (or for that matter to closed buildings), most users are well satisfied. Where they are not it is due to faulty system design, control, or maintenance.

SOURCE AND AVAILABILITY OF ENERGY FOR AIR CONDITIONING

The energy source for summer cooling is almost invariably electricity and is thus independent of the changing pattern of primary fuel availability in so far as system design and control is concerned. Operating costs for cooling are

8

thus almost wholly dependent on the cost and availability of electricity. This will be determined by weightier considerations than the use of electricity for air conditioning. Electricity will form an increasingly large element in the cost of living, including eventually transport. So, in a relative sense (relative, that is, to the cost of living) running costs of air conditioning may not change very much over the years. Intensive studies of world energy demand and supply suggest that under the stimulus of rising costs efforts to conserve energy may dramatically lower the trend lines predicted in 1973-5 while the same stimulus may intensify discovery and exploitation of usable reserves of fossil fuels. So the dreaded day when fossil fuels run out may be postponed for rather longer than we feared.

Some suggest (1) that we may expect to enter the 21st century with two-thirds of our total oil reserves remaining, though of course we will be consuming at a much higher rate than today. This refers to the world, but the output from existing discoveries in the North Sea will taper off dramatically in the mid-1980's, though the oil men are still optimistic that further discoveries will extend output until around 2050. Increased coal production may fill some, but not all of the gap.

So we have a little time, but not too much, for the time required to develop new sources of energy is measured in decades, not years. The more we conserve energy now the more time we shall have to adjust. But here we run into the problem of motivation for as *The Working Document on Energy Policy* by the Energy Commission (2) points out, 'There are also problems of timing. The prospect of higher prices and of energy scarcity ten or fifteen years hence provides only a weak stimulus to the individual to make early investment. His preferred course would be to delay investment until the price increase was imminent, or had actually occurred'.

THE SIGNIFICANCE OF AIR CONDITIONING TO NATIONAL ENERGY CONSUMPTION

In 1972 BRE estimated that engineering services in buildings account for about 45% of national primary energy consumption. (By way of comparison transport only accounts for about 13%.)

Information on the installed capacity of existing air conditioning plants is sparse. Hazarding a rather wild guess, it may be of the order of 1,000,000 tons refrigeration (a less confusing way of referring to it than in megawatts). If this is of the right order, then the total consumption of these plants when compared to the estimates of national primary fuel consumption (amounting to 340 MTCE) leads to the conclusion that existing air conditioning plants account for only 0.25% of national energy consumption, or 0.5% of the energy used in buildings.

Therefore air conditioning makes no significant impact on national energy consumption.

Table 1.1

Energy consumption per annum for a family of four people, one of whom works in an office	Site energy	MJ	Primary energy
Home heating (3 bedroom detached house)	72,000		75,000
Home hot water service	25,000		26,000
Cooking (gas)	11,000		11,000
Lighting, TV, etc	12,000		44,000
	120,000		156,000
Motor car (1600 cc)	38,000		42,000
Totals for 4 people	158,000		198,000
Office heating	4700		5000
Office hot water service	900		1000
	5600		6000
Office lighting	1700		6100
	7300		12,100
Or, alternatively, Office air conditioning (electricity and gas)	7600		12,600
Office hot water service	900		1000
	8500		13,600
Office lighting	1700		6100
	10,200		19,700

Table 1.2

Comparative costs of heating, or heating and ventilating or air conditioning an office and relationship of staff salaries	Heating	Heating and ventilating	Air conditioning
Capital costs £/m²	13.4	47.9	71.2
Annual costs £/m²			
Depreciation of capital	1.6	5.6	8.4
Maintenance	3.3	4.2	5.0
Fuel and electricity	0.9	1.2	2.5
Totals	5.8	11.0	15.9
Total as % of staff salary	1.0%	1.9%	2.7%
Total as % of total cost of employment	0.5%	1.0%	1.4%

ENERGY CONSUMPTION AND COST OF OWNING AND OPERATING AIR CONDITIONING IN OFFICES

Many years ago I prepared some figures to put the costs of air conditioning in offices into perspective by relating them to the costs of employing the staff. It is also of interest to study the individual employee's other uses for energy, such as heating his home or running his motor car. Mr W. P. Jones has kindly reworked these ratios for me. Tables 1.1 and 1.2 show the result. They are based on a three bedroom detached house, a 1600 cc motor car, and an average staff salary of £5300 pa which gives a total cost of employment of £10,000. Fuller details are given in the appendix.

Table 1.1 shows that the primary energy required by a man in a non-air conditioned office is only 6% of that required to service his home and motor car, while if his office is air conditioned it only consumes another 4%. If you regard air conditioning as a heavy energy consumer, then have a look at his motor car. The additional energy required to provide air conditioning for him at work as compared with heating only is only 7½% of what he chooses to use in his motor car.

It is also interesting, and rather surprising, to note how very much more energy is required to service his home. This is explained partly because his home is occupied for twice as many hours as the office, but mainly because the space used by his four member family at home is 10 times as much as he occupies in the office. This serves to emphasise that for office workers, at any rate, by far the greater potential for national energy saving is in the home. For factory workers this would be somewhat less marked because the heated volume per worker in the factory is usually much greater than that in an office (though the factory is unlikely to be air conditioned).

Employers will, of course, be interested in the operating costs of providing heating, or heating and ventilation, or air conditioning, in the office. Table 1.2 gives a breakdown of costs and relates these to salaries and to total employment cost. It shows that to pay for the extra cost of air conditioning an employee's productivity would have to improve by only 0.4%. So it is reasonable to assume that employers will consider it economic to air condition offices, and not only in adverse locations. It would be wrong in terms of national productivity to dissuade them from doing so.

MEANS OF PROMOTING ENERGY CONSERVATION

What has been said so far demonstrates that in terms of national energy consumption air conditioning need not be singled out for special attention. That does not mean that there is nothing we can or should do. Far from it. All energy saving is important. So let us examine the measures that can be taken to save energy used by building services, whatever form the services take, air conditioning included.

Perhaps the most important point to note is that the policies adopted for improving existing buildings would make a much greater and more immediate impact on energy consumption than those for new buildings, since we only add one or two per cent to our new building stock each year. This is, however, a more complex subject than appears at first sight, not least because the end result of improving efficiency, whether in the building itself (eg insulation) or in the heating system, if any, may be to leave total energy consumption where it was before and lead only to raising standards of comfort. Socially desirable, of course, but sometimes counterproductive in terms of energy saving. In fact, it would really be much better to persuade people to wear thick underclothes instead of working in their shirt sleeves!

Designs are likely to be economic in their use of energy if they are based on the long-term costs of owning and operating, especially if energy costs are priced realistically. There is a comment on this (2) in *The Working Document on Energy Policy,* which says, 'The principle that prices should reflect costs of supply on a continuing basis, while providing an adequate return on investment, is now firmly established though problems will inevitably arise in its application'. This does at least give hope of a more rational approach to economical design and, indeed, if the policy had been applied for the last 20 years it would have prevented some of the distortions that have occurred in the choice of energy sources for heating systems.

As to the means by which economies can be encouraged, there is no doubt that price of energy is the most potent motivator, though as has been noted earlier it has limitations in that the economic horizons of building owners or occupiers tend to be too close and seldom go beyond 20% of the life of the building. Nevertheless, realism demands that this tendency should not be ignored. Probably the best course of action is for designers to make better provision for subsequent modification of plant in later years.

Although energy price is the quickest and most effective short-term motivator for change, we live in times when influence of public opinion is an essential spur to action. There is little doubt that the fairly imminent decline in availability of fossil fuels is likely to be the most cataclysmic event of our age and will soon be recognised as such.

The importance of accelerating research and development of new sources of energy has been recognised by governments throughout the world, though whether the scale of effort is sufficient is difficult to judge at this moment. Undoubtedly we have to buy time by energy conservation. The influence of conferences in the moulding of professional and user opinion is an essential precursor of the demands which will be placed on the industry as awareness of our future predicament grows in the public mind. The greater the awareness, the less the need for government intervention, whether in the form of taxation incentives (in any case a blunt instrument, as the Energy Commission points out), or regulations, but bearing in mind the rather short-term effect of price motivation there is little doubt that some form of regulation will be demanded to supplement private effort.

Now regulations can be of two essentially different types – prescriptive or functional, ie detailed requirements for a multiplicity of components or general performance standards. The subject is discussed in the editorial of the December issue of the *Building Services Engineer* (3), and also in an article by A. Newton in the same issue. He says, 'Although the use of controls and other necessary measures could be ensured by prescriptive legislation, energy budgeting is to be preferred. This is because it is a system of limiting energy use in a rational and flexible fashion so it will automatically select the most cost effective measures as they are developed in the future. On the other hand, prescriptive regulation offers savings that are considerably less cost effective because of its restricted scope and its inherent dependence on established techniques in a field that is rapidly changing'.

The CIBS have produced part 1 of their *Building Energy Code* (4) and parts 2, 3 and 4 will follow soon. No doubt Mr Peach will enlarge on this subject in his paper, but let me say here that this Code is a major contribution to the subject of energy conservation. The Code takes the designer through the logical sequence of building form including glazing and natural lighting, insulation, heating, air conditioning, artificial lighting and their control systems. It makes a number of judgements on currently accepted standards which have been developed with energy conservation in mind. But as the Code points out, 'Where standards and targets are suggested they are necessarily from a present-day point of view and their implementation may acquire more urgency and rigour as time passes'. All these things underline the importance of ensuring that any regulations which it is decided to apply are as flexible as is consistent with exercising an authoritative influence.

Technology will advance, more efficient equipment will be designed, relative costs will change, and legislation may itself change for strategic or political reasons. Any prescriptive detailed regulations would be out of date almost as soon as they had been promulgated. The only sensible way forward must be for all, government included, to endorse the CIBS Code, to require engineers to apply it, and to ensure that it is regularly updated in the light of developments. Government and services engineers must act in partnership and private initiative must be supplemented – not stifled.

LOW ENERGY CONSUMPTION BY AIR CONDITIONING SYSTEMS

There are a number of interesting problems to be resolved, not least the optimisation of glazing both in air conditioned and non-air-conditioned buildings. Although glazing imposes a big load on air conditioning systems in summer and on heating in winter, it reduces the artificial lighting consumption, and by admission of heat from the sun in winter provides appreciable compensation for the high heat loss. It provides in effect a solar heating and lighting system and its shape and configuration can be optimised to provide minimum annual energy consumption.

Artificial lighting can also be an important element in cooling loads and the present acceptance of more modest levels than were being suggested five years ago will undoubtedly save energy. The development of task lighting is another matter. Certainly it provides scope for saving but it may also introduce restraints on location of work places which, by their inflexibility, are in opposition to the concept of long life, loose fit buildings.

The use of systems which collect unwanted heat at source and re-use it in other parts of the building where there is a demand for it is an art rather than a technological advance. The art is progressing but needs more practice and development. Papers given to the Institution assist in this process by allowing us to share experience. Heat reclaim methods such as from exhaust air, though quite widely practised, will need further technological development if they are to be economically more attractive.

But all this relates to new systems in new buildings. Naturally, the primary interest of designers is in new buildings but when it comes to energy conservation, though initial design is important, maintenance of operational efficiency is paramount. If building occupants are uncomfortable they will say so and eventually something will be done about it, but if running costs are high it is unlikely that the building owner – already punch drunk with fuel cost inflation – will do anything about it.

This is why parts 2 and 4 of the CIBS Energy Code, soon to be published, will be so important. If the user knows, or can be advised on what is an economical target for energy consumption in his building, then if that target is not met he will react.

The use of targets will itself require development, for if our actions are to be timely they will need to take account of seasonal variations in the external climate and the effects which they may have on consumption. Indeed, there may in future be scope for the issue of regular, perhaps quarterly, bulletins by the Department of Energy in co-operation with the Meteorological Office giving the most recent weather information in a suitable form for application to energy targets. It would not be too difficult to do this for heating, eg by factors related to regional degree data experience, though correction for sun and wind might be rather tendentious. For air conditioning, it would probably be much more difficult and bearing in mind the small part that air conditioning plays in national consumption it is unlikely that figures would be published other than in the normal forms, and these are usually too late to be useful. However, in view of the importance to individual operators it is possible that one might develop a co-operative information service for areas of high use, like London.

In passing it is perhaps worth noting that we still have not come to terms with the problems of office cleaning in the hours of darkness, and the high lighting costs it engenders.

There is undoubtedly scope for the development of more sophisticated maintenance of air conditioning whether by internal organisation within large enterprises, or by specialist maintenance companies. They should deal not only with fault finding and basic maintenance, but also interest themselves in running costs and with the use and maintenance of systematic or computerised data logging and performance optimising control systems which are now on the market.

Quite apart from the importance of keeping the new series of installations in trim, the potential for energy saving arising from a thorough examination and overhaul of existing installations is enormous. We have seen many examples. T. Gray refers to one in the December issue of the *Building Service Engineer* (5). Mr Dubin showed us over an installation in Hartford – the Connecticut Insurance Corporation – at the Institution's summer meeting at Boston in 1975, and many other examples have been publicised.

Getting the best out of what exists is one matter but we must not shrink from major modification or replacement of outdated or badly designed systems, for we must never lose sight of the supreme importance of improving performance on the existing stock of buildings.

REFERENCES

1 *Investments in Tomorrow,* SRI International, vol 7, no 3, 1977
2 Energy Commission Paper no 1, *Working Document on Energy Policy*
3 A. Newton, Energy Budgeting Legislation for Energy Conservation, *Building Services Engineer,* vol 45, no 9, December 1977
4 CIBS, *Building Energy Code,* part 1
5 Tony Gray, Management Opportunities in Energy Conservation, *Building Services Engineer,* vol 45, no 9, December 1977

APPENDIX: Data used for compiling Tables 1.1 and 1.2

House	3 bedrooms 90 m² total floor area LTHW heating by gas fired boiler
Office	Typical 14,000 m² block with 50% of long sides single glazed Space per occupant 9 m² Gas fired boiler Perimeter induction air conditioning 1.3 litres/s/m² fresh air 25 W/m² lighting
Car	1600 cc engine 26 mpg 9 000 miles per annum

Employment
Average staff salary £5 300
Total cost per person in London including holiday, insurance, pension, rent, rates, fuel, etc £10 000.

Cost of system
Heating by LTHW finned tube in casing or air conditioning by perimeter induction.
Electric wiring included for plant items.
(H & V) ventilation option 5.2 litres/s/m²
20 year system life.
10% interest on capital.

2 Comfort experience

E. C. LOVELOCK

INTRODUCTION

A considerable amount of data is available in textbooks and codes of practice which very meticulously details the criteria for designing an ideal interior climate for a building. The title of this chapter calls for the reaction to this design criteria and puts the topic in a very subjective area: no two persons are alike in their sensitivities and no two buildings perform in the same manner. To narrow the subject down into manageable proportions the subject is kept to office premises in the UK.

Quantification of the subjective becomes rather difficult other than in general statements because little meaningful data is available. These statements are based on impressions gained through experience and discussion with a number of building managers. They are not in themselves finite but if they promote further discussion some refinement may be possible. Frequent reference is made to Shell Centre where records have been kept which substantiate the statements made by a number of building managers. Shell Centre is the Central Office of the Royal Dutch Shell Group of Companies in London. It has a gross floor area of 175,029 sq metres and is air conditioned throughout. In addition to offices, accommodating a working population of approximately 5000 persons, it has extensive recreational and entertainment facilities.

Although this conference is restricted to air conditioning as its main topic one must question whether the building occupants are able to separate out their reaction to thermal comfort from the total environment experienced. Buildings in the UK, having air conditioning, tend to be rather different in concept (constructed in circumstances which give the requirement for the more sophisticated services) from the non-air conditioned building. The former may be an open landscaped office or high rise with high glazing ratios whilst the latter may be a traditional building partitioned to provide offices for a low level multi-occupancy, the cube per person may be higher and of course the window design will be different.

A building owner or occupier could make serious errors of judgement if he views air conditioning in isolation and not part of a total package. It may be an inappropriate assessment of priorities if he seeks ways of reducing costs with a consequential reduction in comfort if such a course of action is not in context of the total built environment and useage of energy. The occupants would feel incensed if comfort conditions are sacrificed in the

16

course of energy conservation and at the same time office lighting is left burning throughout the night when the building is left unoccupied. By way of introduction it may therefore be helpful to set air conditioning firstly in the total system of building operation and then subsequently the overall energy consumption.

A further factor when assessing comfort conditions is that as mentioned above opinion of what is good, bad or indifferent is very subjective. A complaint about thermal comfort may well form a tangible way of expressing unhappiness about a situation which is disconnected but has little outlet other than a moan about something as a form of relieving a stress condition. How often do we in the UK attribute a lack in the sense of feeling of wellbeing to the climate?

THE BUILT ENVIRONMENT

When financing a development careful consideration is given to the capital involved and this will determine the degree of sophistication built into the project. With electrical and mechanical services representing a large proportion of costs, possibly as high as 50% for specialised buildings eg hospitals, computer suites etc, there will often be difficult decisions to make, when trimming the cost plan, with regard to where cuts should be made. It is not always appreciated that whilst a reduction in finishing standards can be restored at a later date a reduction in the services, which have a direct bearing on the environment will be difficult if not impossible to restore once the installation is completed. Often the problems experienced in a building environment reflect a cost cutting exercise and the aggravations so caused to the occupants, which will be a demotivator, may outweigh the money saved. An unsatisfactory air conditioning system may well be worse from the point of view of gaining acceptance by the occupants than cutting back to a straightforward central heating system with mechanical ventilation associated with opening window sections.

In these days of industrial democracy, most companies, other than the comparatively small, have staff representation either by union or staff committees. Such representations are paying more and more attention to environmental problems. It has often been said that the best that one can achieve in air conditioning is to satisfy 80% of the occupants but it may be forgotten that it is only the complaints that one hears and the remaining 20% can be extremely vociferous.

It is important that there should be a clear understanding of the standards of environment to be provided and aspirations should not be unduly raised above that which can be realistically provided within both physical and cost constraints.

When justifying capital expenditure and the return on investment it will be necessary to consider the life cycle costs. Whilst initial building costs are well understood and considered when carrying out a development the same depth of understanding may not apply to the operating costs when in occupation. Fig 2.1 represents the make up of the annual costs for owning and occupying a large complex (Shell Centre).

Whilst the operating and maintenance costs may in themselves be considerable sums of money, the percentage of the total presents a vastly different picture. If the services costs are again broken down (fig 2.2) it will be seen that the maintenance of air conditioning represents 38% of the total.

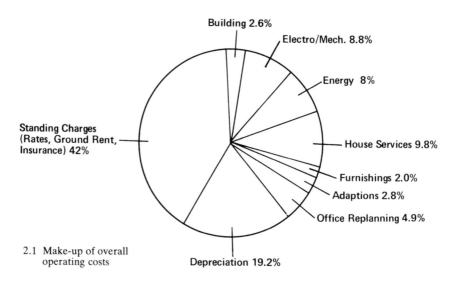

2.1 Make-up of overall operating costs

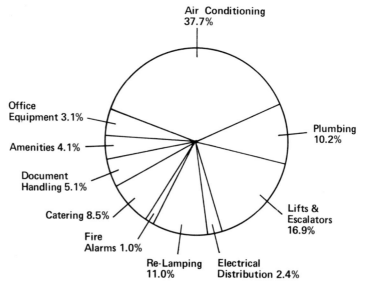

2.2 Make-up of maintenance costs

The purpose of putting forward the above data is to get a full perspective on the costs involved in the environmental package. The building manager should be fully aware of the operating costs. From time to time he will be under pressure to make savings and it is important that he should be in a position to fully appraise the total situation and make savings in those areas which cause least hardship. Too often it is assumed that air conditioning is the only real factor making up maintenance costs and when making savings this is the single target on which to concentrate. The result could be a plant failure, for which there is no standby and loss of control. Such consequences become evident fairly quickly in terms of a deteriorating environment but take a long time to correct. So far as the occupants are concerned there is a lack of confidence in the system and sensitivities become heightened. Perhaps the savings could be made in other areas without the same effect on the environment.

It is vital that the built environment once installed and properly set up be maintained in that condition, or if changes are to be made there is a logical approach with due warning rather than allow it to happen in random fashion causing loss of control. Over a period of time some areas within the building may suffer a gradual change in use. This will cause complaints about the environment and much time will be wasted in the investigation and making ad hoc changes if it is not recognised that there has been a deviation from the designer's brief.

When a building is handed over to a client the operating and maintenance manuals should include design parameters eg occupancy factors for offices, conference rooms and public rooms to give an indication when they are being overcrowded, the designed heat loads allowable from portable office machines etc. Office machines, with a greater emphasis on mechanisation, give frequent cause for complaints arising from local hot spots. It is important to establish a liaison contact with those responsible for ordering and installing office machines, especially photocopiers which can have electrical loading of several kilowatts, so that any need for re-balancing or alteration to the air conditioning system may be carried out in advance of the machine arriving.

Packaged cooling units, using either water or building exhaust air for condenser cooling are very useful for office areas carrying a very heavy heat load above design criteria. The packaged unit controls may be set inside a range of the standard control system and so superimpose additional cooling capacity within fine limits of $\pm$ 1°C dry bulb and $\pm$ 5% rh, which may be required for computer based hardware.

THE THERMAL ENVIRONMENT

The *IHVE Guide Book A* gives very detailed information on the parameters for designing the environment. A typical design of internal climate of an office would be:

Winter	external : $-2°C$: 80% rh
	internal : 21°C : 50% rh
Summer	external : 26.5°C : 54% rh
	internal : 23°C : 50% rh

Note

Tolerance on:	internal temperature $\pm$ 2°C
	relative humidity $\pm$ 10%

At first sight the above tolerances appear to be very precise especially when related to central heating systems with radiation surfaces only or the conditions in one's home. In actual practice the body becomes very tuned to an air conditioned environment and such limits are easily sensed.

Experience indicates that for winter conditions a minimum dry bulb temperature of 20°C for sedentary workers is barely acceptable for males and will bring protest from females. This has been amply illustrated by the reaction to the 1973 Government Order, still in force, restricting internal temperatures in offices to a maximum of initially 17°C now raised to 20°C (68°F). It is noticeable that this restriction has brought about a change in dress habits, heavier suiting for men and the popularity of trouser suits for women which is leading to a greater acceptance of the lower temperatures.

During the summer, temperatures above 24°C start to be uncomfortable, especially in areas where the work is less sedentary eg involves a fair amount of movement between work stations such as in mailing, teleprinter suites etc. It would appear that the higher temperatures become intolerable at a lower order and far more quickly with air conditioning than in a naturally ventilated uncooled environment. Possibly this is due to the fact that the occupants are unable to do very much about it in an air conditioned building whereas in a centrally heated building they are able to open windows and spend a little time near them to relieve their stress. However, summer conditions give rise to far less complaints than winter and certainly when the outside conditions are in the upper regions of say 27°C+ and there is little intrusion of noise due to sealed double glazed windows the air conditioning systems are most appreciated.

Experience indicates that the preferred average office temperature seems to be a constant 22°C with a rh of not less than 50% all year round with the ability of slight adjustment by thermostats. Dress habits change to suit this constant environment and the same weight clothing is worn throughout the year putting on heavier clothing in the winter only when leaving the office. It should however, be emphasised that 22°C is above that permitted by the 1973 Government Order during the heating season.

It is stated in Codes of Practice that when entering a cool building from an external high temperature summer condition, thermal shock will be experienced if the differential exceeds 5°C. The suggested constant all year round temperature of 22°C is contrary to this statement when external temperatures exceed 27°C. However during hot periods of 30°C+ there has been no suggestion of thermal shock being experienced, internal humidity under these conditions would probably be approaching 60% rh. The preferred constant temperature is also contrary to the belief held by some that such conditions are soporific and a changing environment is necessary to create stimuli.

Air movement, the third member of the comfort index is far less tangible than temperature and humidity. Very little air movement gives a feeling of stuffiness and the air can be smoke laden whilst too much will cause a feeling of coldness far more quickly than a lowering of ambient temperature.

Uncomfortable air movement will be caused by air moving at a lower temperature than the ambient. It is considered that air movement of 0.125 m/s constitutes still air and will normally appertain without fans due to normal convection currents. However under draught conditions in air conditioning eg work stations close to windows, air movements as low as 0.025 m/s may be felt on exposed sensitive skin such as on the back of the neck. For sedentary workers 0.125 m/s appears to be the upper limit which is acceptable for air at about the ambient temperature and women appear to be more sensitive to air movement than men.

For more active work stations and particularly in those cases when operating heat producing office machines the obverse to draught conditions exist and some directional air movement will be welcome to dissipate the heat. In public rooms, high air movement is intolerable when sparsely occupied and most welcomed when crowded which makes for an impossible situation unless variable speed fans or a variable volume control system are used.

It is probable that air movement is one of the factors which causes a transformation in the character of persons moving from a centrally heated environment where very poor standards of thermal comfort are tolerated to an air conditioned environment where very high standards are maintained within good tolerance and yet the slightest deviation gives rise to deeply felt complaints.

During winter conditions the higher movement of air with air conditioning equates to a comfort index some 2°C higher than in a centrally heated environment eg 20°C may be more acceptable in a centrally heated building where there is little air movement than in an air conditioned building, and the possibility of sitting near radiators.

Air movement will also have a bearing on the comfort condition for humidity. Again in a centrally heated building 35% rh appears to be acceptable in the winter whilst in air conditioning a feeling of dryness starts to be felt below 50% rh. Further comments on these factors are given under the heading below 'Commissioning and Acclimatisation'.

It is suggested that insufficient consideration is given to air movement at the design stage: variable blades in air inlets, whilst a great help, do not provide the complete answer particularly in those systems requiring a long throw of air, initially at high velocity, to ensure full circulation. Far more work should be carried out on 'mock-ups' of rooms, using smoke to trace the air movement. Perhaps more work at the design stage would eliminate the need for occupants to apply to a very sophisticated, finely commissioned and balanced system, pieces of paper over grills to reduce air movement or streamers to prove that some does exist!

A further factor which is difficult to correct is radiation from cold surfaces. With the ideal conditions of a constant 22°C these problems are less evident than when operating with the permitted lower maximum of 20°C, such problems can make some work stations intolerable and a change in office planning may inevitably be necessary to remove people who have their backs to such cold surfaces.

Typical problem areas are walls enclosing stairs. Such staircases often have inadequate convector heating surfaces designed on the basis that a comfortable temperature is not necessary for negotiating stairs. If they are a means of escape route the problem is further aggravated by smoke vents. This condition may be mitigated by the provision of shutters over the vents which will fall open when activated by a mechanism triggered by a smoke detector. The coldness of such wall surfaces with consequential radiation is felt by those in adjoining areas. Further examples are perimeter walls on exposed aspects and floor slabs which are exposed to the elements on the underside with little insulation eg over external walkways, parts of the building raised on columns to permit traffic movement at ground level or overhang building at first floor levels. If these exposed sections are critical, and they should be examined at design stage, then suitable heating elements or insulation are best designed into the system. Finding remedies at a later stage by adding insulation materials or electrically heated carpet underlays is only a partial solution.

Troublesome cold surfaces in buildings may easily be detected by infra-red photography which produces thermogram prints varying in colours, the colour being a fairly accurate indication of the surface temperature.

Badly designed entrance halls having inadequate draught exclusion can be a source of annoyance. The traditional revolving door is no longer a solution to draught free halls by present day standards, particularly on exposed sites. In this context one should bear in mind that an apparently troublefree site undergoes a change when it is fully developed and surrounded by other buildings which due to their placement and height cause high winds to be experienced. Well designed draught screens with some form of hot air curtain are required if the entrance is to be comfortable for the security/ reception staff who have to work there and to prevent cold air circulation through adjoining areas.

Problems associated with solar direct or indirect radiation are usually well-considered at the design stage. It is an element which is fully assessed when calculating cooling loads. Solar heating should not be a comfort problem in well insulated buildings having a reasonable glazing ratio say 25% total wall area. Where sun blinds have been provided in such circumstances they are rarely used by the occupants who prefer the uplift of the sun shining into their offices.

Although not a direct factor of the comfort index, well filtered air associated with closed windows improves the built environment. The benefits are not quite as evident today with Clean Air Acts giving a reduction in carbon particles in the atmosphere. However, the extended periods between cleaning or redecorating interior finishes points to the cleaner air circulated in the building. Filtration will also remove pollen which should be of help to those who suffer from hay fever. In densely developed areas high level air intakes plus filtration will eliminate obnoxious odours and reduce traffic fumes. The closed windows will also reduce the intrusion of high level external ambient noise.

A stable environment meeting the criteria given above will, it has been proved, provide a high standard of comfort. Unfortunately few building managers keep accurate records of complaints but the data relative to Shell Centre may be considered typical. The method of making a complaint is made as easy as possible so that remedial action may be carried out without delay rather than allow the aggravation to accelerate. Any member of the staff may telephone a complaint to a central logging station.

Fig 2.3 indicates the complaints made for the years 1972 (prior to the restriction on permitted temperature) and 1976, shown month by month. The total number of complaints for 1972 and 1976 were 731 and 767 respectively. In 80% of these cases a fault was found in the modular control system leaving only 20% of the complaints being real dissatisfaction with the average comfort conditions provided. These roughly fall into the following categories.

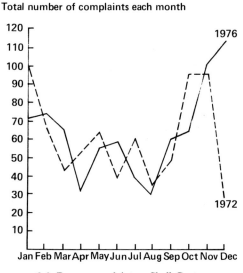

2.3 Room complaints – Shell Centre

- 12% antagonistic towards Government Order restricting temperatures to 20°C
- 2% in multi-occupancy offices where a different condition from the average was requested
- 2% from staff on flexi-time and starting work before design conditions reached
- 2% from small pressure group where in fact only one faulty module was found
- 2% unspecified problems which resolved themselves.

Average complaints are less than four per day or 1·5% of total room modules.

The higher level of complaints during the winter of 1976 compared with 1972 indicates that the lower set temperature searches out the cold spots plus some resistance to the lower temperature. All rooms are provided with thermostats and a winter survey of the setting indicated 10% below 20°C; 60% were at 20°C and 30% were set above 20°C. Considerable efforts have been conducted in energy saving campaigns and although some initial resistance was evident, co-operation has improved and there has been a change in attitudes. Needless to say that at this time the higher temperature of 22°C in winter would be preferred and found more comfortable.

The stable environment however, can only be obtained if the building is well designed with good mass and level of thermal insulation to prevent violent temperature swings from outside conditions and the plant well maintained.

Ideally the designed environment should be met from the first day of occupying a new building and for the whole time it is occupied, from the start of work on Monday morning through the week. Unfortunately these conditions are seldom if ever met which is the cause of much dissatisfaction with air conditioning. It is necessary to go through a period of commissioning and acclimatisation of the occupants – a period of two to three years – before approaching an acceptable stable condition. Day to day fluctuation could be caused by the need to economise on costs so leaving the start-up as late as possible in the early mornings without sufficient anticipation of outside conditions and secondly control systems which are not sufficiently responsive in anticipating outside conditions. As mentioned flexi-time working does not help in this connection.

COMMISSIONING AND ACCLIMATISATION

The first year of moving into a newly completed building provided with air conditioning can be extremely traumatic for the occupants. Often the plant gives a very bad standard of performance due to teething problems whilst the staff themselves are being subjected to a change in habit by moving into a new environment, possibly in a different location with travel problems and it is not surprising that at times tempers rise near to break point.

It is usually stated by the installers that one year is required to balance an air conditioning system. Unfortunately the term balancing sometimes includes a multitude of incomplete items in addition to commissioning etc and it is not unknown for a building to be occupied whilst the services have only a rudimentary control system in operation. This may be due to the installation and commissioning work either slipping in the construction programme or not being phased in sufficiently early. All the care and commitments exercised by the designers is completely negated at this stage and the occupants may become so aggrieved with heightened sensitivities that the situation can get out of control. Furthermore these first impressions live for a very long time with the consequence that full confidence in the air conditioning is not established for several years.

One would think that when moving into a sophisticated air conditioned office the persons concerned would be highly delighted with what they find. The anomaly is that the reverse is the case which is perhaps difficult to understand. Whilst psychologists have noted this apparent anomaly the real cause has not been explored. After a period of adjustment however, the controlled environment is appreciated.

When moving from a controlled environment to non-air conditioned, dissatisfaction is first experienced but the period of adjustment seems less than the reverse movement except when the external environment is poor in terms of noise and fumes or during short summer spells when high temperatures are experienced. Perhaps this would not be equally true if air conditioning was more prevalent and became the norm.

Quite recently a case was reported in the daily press of a new building costing some 14-15 million pounds being occupied and the occupants were so unhappy with the environment that a strike was threatened. It was claimed that conditions were so bad as to cause attacks of bronchitis, fibrositis and yet the anomaly is that the thermal comfort was probably superior in real terms to that previously experienced.

It is apparent that when working in an air conditioned environment for the first time, a period of acclimatisation is required to become accustomed to the new climate over which there is little scope, if any, for adjustment by the individual. In the case of a new installation this process is required to take place whilst the system is not properly balanced and commissioned. During this period of both the occupants and the installers coming to terms with the problems of adjustment, it is essential for frank and honest discussions to be conducted so that the occupants may be kept fully up to date with what is happening and the remedial action being taken. If the occupants are able to understand the problems facing the installers and be aware that they may have to change their attitudes towards weight of clothing worn etc to gain the full benefit of air conditioning then it may be possible to ensure a smoother transition from first occupation through those first two or three years.

It is however, suggested that there is more scope for research into the techniques of balancing and commissioning systems and methods of programming this work such that it may be commenced at the earliest possible date in the programme. It is appreciated that this is a cost element in the contract sum and earlier commissioning may be more costly than leaving until a later stage but at least the client should be given the option of spending more money if he is to save much of the aggravation which will be caused when occupying the building. It is interesting to note that when installing services for computer suites the system is proved by simulated heat loads before handling over to the client, possibly packaged units complete with pre-set control systems may be of help in this connection. Could more of this technology with modular partial commissioning at an early stage help to reduce all the teething problems experienced?

A case study on the occupation of Shell Centre is typical of the problems met on first occupying a large complex. Whilst this happened some years ago the events still appear valid at the present day.

The first reaction on occupying a temperature controlled environment is to experiment with thermostat settings. The thermostats provided had a dial calibrated 13-29·5°C which gave the impression that any temperature within this range is obtainable. The person wishing a cool environment would turn the thermostat to the lowest position which in the winter results in a temperature approaching 20°C. However, on a cold Monday morning after the shutdown over the weekend a very much lower temperature would be found due to the thermostats holding shut all the heating valves in a particular module. This gave rise to complaints on Monday mornings and only after considerable investigation was it found that the problem was mainly due to a misunderstanding on the operation of the thermostats.

It was generally found that the female staff required a temperature above 21°C and were very aware of draughts whilst the male staff wished for temperatures closer to 20°C with air movement. Over a period of several years this preferred temperature crept up about 2°C which perhaps cured some of the problem areas where for a variety of reasons it was not possible to reach the average temperature during the winter. Of the two factors temperature and air movement the latter proved to be the most significant. It was surprising to find that the staff who had moved from centrally heated buildings, some of which were very old and gave a very poor standard of environment, quickly became sensitive to small changes in temperature and a difference of 1°C became critical.

It is probable that the type of clothing worn at that time had some influence on comfort factors. The tendency was to wear heavy suiting and woollen garments in winter and change to much lighter clothing in the summer yet internal conditions were virtually constant the whole year round. An understanding of these factors led to more comfortable weight of clothing enabling the full benefit of air conditioning to be appreciated.

A mean dry bulb temperature of 22°C was adopted and found to be the average requirement for sedentary workers. In areas, however, where more physical work was involved a slightly lower temperature of say one or two degrees would have been more acceptable but such variations were difficult to achieve within the control system. Whilst wide variation of temperatures to meet individual requirements was impracticable adjustment of air movement solved some problems. Fortunately the type of air vent grilles fitted provided easy adjustment of direction and intensity of the air movement in local areas and some re-balancing proved necessary. The installers provided full records of air-balancing and changes have been noted to ensure that overall performance is maintained.

In the early days, concern was expressed regarding the dryness of the environment – the combination of air movement and humidity. Some female staff complained of dryness to the skin and the medical unit were under the impression that sinus complaints may have been on the increase. Data were not however available to quantify these comments. It was decided to commission a study by an ear, nose and throat consultant to establish whether there was any significant difference in these respects for air conditioned and for centrally heated offices. Shell Centre was compared with two large centrally heated offices in the immediate vicinity with the following results:

1. 990 questionnaires were sent out for Shell Centre, 450 for one of the centrally heated offices and 610 for the other
2. The incidence of upper respiratory tract infections was similar in air conditioned and in centrally heated offices. Similarly, using the incidence of illness lasting seven days or more there was no difference in this respect
3. The percentage of occupants who expressed satisfaction with comfort conditions was:

	Air conditioned (Shell Centre)	% Satisfaction Centrally heated Office A	Office B
Winter			
Heating	78·8	66·5	63·6
Humidity	73·5	68·2	63·6
Air movement	63·5	46·6	36·5
Summer			
Heating	77·4	78·7	80·5
Humidity	73·4	81·7	79·8
Air movement	62·8	64·4	60·6

4. The percentage of occupants finding greater comfort in their present rooms compared with those previously occupied was:

Air conditioned (Shell Centre)	69·4%
Centrally heated Office A	57·8%
Office B	52·8%

The main object of the study was to determine whether air conditioning produced nasal symptoms to a greater degree than a centrally heated environment. The fact that no difference was found gave a satisfactory answer. The findings regarding comfort conditions are rather problematical, bearing in mind that once one set of occupants had become accustomed to the higher standard of air conditioning they had at the same time become more critical of their environment.

A large proportion of the dissatisfied occupants in all three offices showed a preference for a higher humidity in the winter. In the summer the same comments applied to Shell Centre but in the case of the centrally heated offices the greater percentage was for less humidity.

It was of interest to note that of all three parameters (heating, humidity and air movement) the last proved to be the greatest cause for complaint. This was less critical in Shell Centre during the winter but more so in the summer. This probably reflected the early attitude to the closed window situation. Fortunately in the case of Shell Centre as mentioned, some variation of air movement was obtainable by adjustment to the air inlet grilles.

Throughout the period of adjustment occupants were encouraged to register their complaints by using the 'MEND' telephone facility. Complaints were checked on the same day that they were registered. By checking the log it was possible to establish whether the pattern indicated a general dissatisfaction due to zone control or an individual peculiarity. Over a period of time, both occupants and plant settled down and the number of complaints per day now averages less than four for some 5000 occupants which must surely be a credit to the designers.

ENERGY CONSERVATION RELATIVE TO COMFORT CONDITIONS

Before concentrating on energy consumption relative to air conditioning it is pertinent to set this time in the total energy pattern for a building. Fig 2.5 indicates the energy consumption for a typical office complex for the years 1964 – 7. It also shows the electrical connected load. It will be noted that this load shows a fairly sharp increase in the earlier years. This is due to an increase in office machines and the installation of computer hardware.

Dealing firstly with the total energy conservation scene the industrial labour problems in 1972-3 brought about a very dramatic urgent need to conserve energy. Government Orders and the nature of the situation caused a change in attitudes. This situation was quickly followed by the world energy crisis and sharp inflation in energy costs together with the realisation that energy resources are not unlimited.

Discussions with building managers have indicated that a reduction of energy consumption of something like 25% since the energy discontinuity in 1973 is readily obtainable. This pattern is similar for both air conditioned and non-air conditioned buildings, without greatly reducing comfort condition other than the acceptance of the slightly lower temperatures in the winter.

Examples of the measures taken to achieve these savings are:

- Installation of more efficient burner and control systems to boilers. Although expenditure is required for these measures a payoff in 2-3 years is obtainable.
- A better orientation of heating zones relative to the exposure of the building faces to the external climatic elements in some cases breaking the zones into smaller units where there is a difference in the occupational times.
- A reduction in the daily heating period by better programming of the morning start-up and evening shut-down times and relating this to weather forecasts.
- In the case of air conditioned buildings a more efficient programming of cooling plant operation similar to above.
- The replacement of tungsten lighting with fluorescent and more recently a secondary stage of substituting with the more efficient fluorescent tubes together with improved luminaires.
- Better control of lighting when buildings are being cleaned and consideration of cleaning during daylight hours.
- Education and encouragemant of staff to be energy conscious with particular reference to light switching and use of hot water.

The above has been seen as a total package with the staff occupying the buildings being equal partners in the campaign. In many aspects they have benefited by improved comfort by better heating control and lighting whilst at the same time saving energy. By good publicity with the use of posters and an interest being shown in their welfare a good spirit of co-operation is obtainable.

As mentioned 25% energy saving on office premises whether air conditioned or not is possible fairly quickly without greatly affecting the environment. Such efforts of sound engineering and practices with good energy housekeeping should be the first target before embarking on more sophisticated approaches which may change the thermal environment.

The second stage of improving on the 25% saving will be seen as further refining and a more radical change in plant design and operation. In broad terms, however, a further breakdown of fig 2.5 may be a useful introduction. The energy consumed for the heating system is slightly higher than the electrical consumption in the typical air conditioned building. The gas consumption used in this case for catering and sundry applications is negligible in the totality.

When comparing a stock of non-air conditioned offices in the London area with those which are air conditioned it was found that the energy used in the heating system per unit of floor area is similar for the two types but electricity consumption is three times greater in the air conditioned office. In analysing this statement it should perhaps be said that the non-air conditioned buildings were probably built in the 1930s whilst the air conditioned were constructed in the 1960s.

The similarity in heating energy for the two types is probably explained by air infiltration in the case of the older non-air conditioned buildings being about the same as the two fresh air changes per hour introduced by air conditioning and in both cases the energy for non-environmental services eg domestic hot water, catering etc being the same.

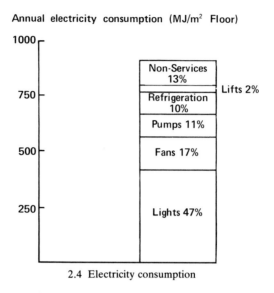

2.4 Electricity consumption

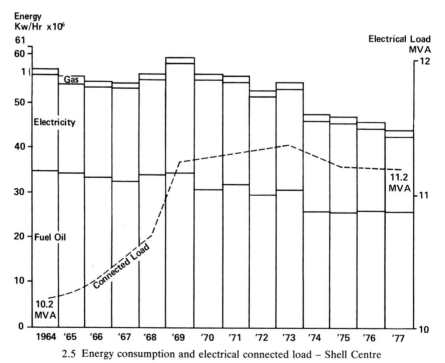

2.5 Energy consumption and electrical connected load – Shell Centre

With regard to electricity consumption a breakdown for air conditioned building is shown in fig 2.4 from which it will be noted that lighting accounts for some 47%. As an absolute figure this is probably greater for the more recently constructed air conditioned buildings having higher levels of illumination. The comparatively high level of energy consumption for plant running points to the need when designing new installations, for a close consideration of overall efficiency in terms of the horsepower required on the driven shaft for fans, pumps, compressors etc and the efficiency of electric motors and associated gear. It will also be of interest to consider the energy required for driving the various types of air conditioning systems eg induction systems versus dual duct.

The energy required for heating or cooling, putting aside the architectural considerations in terms of glazing ratios and insulation, will be determined by fresh air requirements for ventilation and the possibilities of exchanging heat or cooling from exhausted air to that which is drawn into the systems. Thermal wheels for the latter will no doubt start to show attractive payoff periods as energy costs rise. Fresh air requirements to provide the necessary oxygen content; to prevent CO_2 concentrations rising above acceptable limits and the removal of objectional body odours is worthy of further investigation. Current codes of practice have a considerable margin over the absolute requirements.

The above factors could be implemented without affecting standards of comfort and in fact may well show a greater return on energy saving than greatly modifying presently accepted comfort conditions. Bearing in mind however, the energy required to drive the air conditioning plant, consideration is being given to the optimum time for plant operation. Computer based supervisory systems are available for such an approach but very little experience has been obtained in this country on the air conditioning application. The main financial payoff is in the reduction in maximum demand, this saving however, is disproportionate to the energy saving. Trials are now under consideration for shutting down air handling plant for short periods in offices, particularly at mid-day when the occupants are having their lunch break. Whilst it is considered that thermal comfort will not be affected the sudden reduction in ambient noise levels may be a problem particularly where the noise from air inlets has been used to mask other sounds.

Another consideration is the acceptability of temperature swing with outside conditions. Where this has happened it has been due to poor control or faulty plant operation and has been a random pattern which has caused complaints for swings of 2-3°C from design. Such complaints have been associated with buildings of low mass and high glazing ratio and mainly due to solar heating. However, if temperature swings are properly controlled and linked into the supervisory systems mentioned above there may well be scope for gaining acceptability.

Flexi-time systems of working have lengthened plant operating times by some 25% and in theory a similar increase in overall energy for environmental and lighting requirements although in actual practice part of the additional occupation time has been absorbed by cutting down on generous start-up and shut-down times and the acceptance of a slightly lower standard of comfort during the extreme periods. This aspect could however, be more efficiently controlled by a supervisory system.

To conclude it should be said that comfort is that which is perceived by the individual and is therefore greatly affected by attitudes. Similarly the use of energy in offices will be greatly affected by attitudes. However, considerable savings are obtainable without affecting set comfort standards. If attitudes can be further modified then slightly different criteria of comfort will also be accepted which together with energy conservation engineering in design and operation could increase the present obtained overall saving of 25% to something like 50% for air conditioned offices using 1972 as the base.

DISCUSSION

Dr L. Longworth (University of Manchester Institute of Science and Technology)   Mr Lovelock has found that people are more sensitive to change with air conditioning. Does he mean air conditioning at any time of the year or air conditioning when cooling is being used?
E. C. Lovelock (Shell UK Ltd)   My reference to moving people from an air-conditioned building to one without and the acceptability of the two types of building was in the context of air conditioning being provided in its most sophisticated form, ie with cooling and humidification. There are problems in the assessment of the merits of rather subjective judgements particularly as the two types of building are so totally different. An air conditioned building, by the very nature of its design concept, requires a built thermal environment.

In both cases the full environment will have some aspects which are more acceptable than others and when making a comparison there will be a tendency to trade off the bad points against the good. The case study I gave on acclimatisation is based on moving a group of people from a variety of buildings constructed in the 1930s with fairly simple heating systems to one built in the 1960s with full air conditioning. Whilst the air conditioning presented problems during the settling in period it must be said that there were a number of other environmental problems of a more general nature which gave cause for complaint.
J. C. Torrance (Steensen Varming Mulcahy & Partners)   We have experienced problems in a rather large air-conditioned building similar to those referred to by Mr Lovelock with exactly similar reactions, although in our case there was no study by an ear nose and throat specialist. In spite of assurance from the specialist that the nature and incidence of upper respiratory tract infections was similar whether the rh was above 50% and the building air conditioned or below 35% and the building heated Mr

Lovelock said rather positively in his presentation that he would not go below 50% rh. I would suggest 40% might be a better choice as there are very good technical reasons in the context of condensation problems where rh at this level might be helpful.

E. C. Lovelock   My comments on humidity are not based upon a weight of scientific data, indeed they are contrary to tests carried out under laboratory conditions and to the Shell Centre case study to which Mr Torrance referred. They are based on my experience and that of a number of other building managers that if the relative humidity falls below 50% there are complaints about dryness. This is no doubt a product of air movement because in centrally heated buildings the humidity frequently falls below 50% rh without comment.

J. Harrington-Lynn (Department of the Environment)   Mr Lovelock asked for experience of humidity effects on comfort. I worked in a company, which had large areas, kept at 8% rh and 20-25°C air temperature, where girls worked, mainly seated, in assembly lines. Work stopped if the rh went above 10%.

We found that approximately one girl in 100 could not work under these conditions due to medical problems, eg asthma and other chest or throat complaints. These were extreme conditions of course, but the only other effect appeared to be that consumption of liquid was high.

E. C. Lovelock   The very nature of the process in this case dictates the environment. The worker in accepting the job has to adapt his or her mental outlook accordingly. No doubt some form of protective clothing will be worn and the whole work situation is totally different from an office type building accommodating sedentary workers.

Medically, there may not be any justification for humidification on health grounds, but if full consideration is to be given to comfort conditions I submit that very careful thought is necessary before discounting the need for humidification and the consequent small additional expenditure of energy.

D. Arnold (Troup Bywaters & Anders)   I see the tolerances on air conditioning as central to the amount of energy used to maintain thermal comfort. Mr Lovelock suggests allowing a variation of ±2°C which is I believe too great; people cannot be expected to tolerate a temperature variation of 4°C in one day although such a variation might be acceptable over a much longer period. If the temperature can be varied within such a range but with much finer automatic control limits for day to day operation we will maintain better comfort for the individual with lower energy consumption. Such a system might have a variable controlled temperature band, say 21-25°C with ±1°C tolerance around any set point in the band.

E. C. Lovelock   I think your proposal would gain acceptance if the rather tight control you are suggesting could in fact be achieved.

M. N. Carver (Steensen Varming Mulcahy & Partners)   I feel that the expectations of people moving from an old building to a new air conditioned one are too high. On day one perfect conditions are expected from 09.00 hrs in the morning until 17.30 hrs. This is surely unobtainable and possibly not even desirable. I would suggest that conditions be allowed to float perhaps in conjunction with outside air temperature thereby minimising energy use. Savings could again be achieved if humidity were allowed to float too, subject to the constraints of condensation requirements.

E. C. Lovelock   To put temperature swings and their acceptability in context: if on a cold winter Monday morning the external temperature is say 0°C and the internal temperature had fallen to 18°C a swing of 4°C from set-point because the heating is not able to cope, there would be considerable protests from the occupants. Such a temperature swing at that particular time would not be acceptable.

If, however, it is proposed that temperature swings be controlled so that they occur at a time and with a differential such that the conditions are opportune then this is an area worthy of study. Computer based supervisory systems together with controls which are sensitive in anticipating the outside drift in climate and have regard to the thermal characteristics of the building will be of great help in assembling such a system. No doubt these more sophisticated controls will become more economic as energy prices rise.

A. L. Benns (Property Services Agency)   In his paper Mr Lovelock gives a table showing percentage satisfaction. It is intriguing that there is less difference in satisfaction in the summer between the non-air conditioned office and the air conditioned than there is in the winter. This is just the opposite of what would have been expected which would have been that the non-air conditioned office would have been less comfortable in summer—the very reason for air conditioning. This rather surprising fact does suggest that many places are air conditioned unnecessarily.

E. C. Lovelock   The findings of the comfort study should be viewed bearing in mind that one set of occupants have become accustomed to the higher standard of air conditioning and would therefore view this as the norm. In the case of the centrally heated offices the occupants were viewing their type of environment as being the standard.

The people in the air-conditioned offices had heightened their expectations and their sensitivities so that in experiencing a higher standard of comfort they became far more critical. I do not think it would be appropriate to take the statistics given in a literal sense of giving an impression of the acceptability of centrally heated offices versus air conditioned. To provide this kind of information would require a properly controlled scientific investigation.

3 The CIBS Building Energy Code

JACK PEACH

HISTORICAL BACKGROUND

The early 1970s witnessed an increasing realisation that energy is likely to be in short supply sooner rather than later. Whether one projects a usage rate continuing exponentially or constant at the rate at which we entered this decade then fossil fuels will not meet our requirements by the time another quarter-century is passed. Neither is there hope that other parts of the world could come to our aid – if anything, we are better placed than most (fig 3.1).

There are only three possibilities: do nothing, use less, develop alternatives. The first inaction can only lead to chaos, the last action entails an enormous expenditure of time, money and effort – although undoubtedly it is the only reasonable one if we give any thought at all to our children's future well-being. But, for the remainder of my working life at least, the middle option is a must for avoiding disaster.

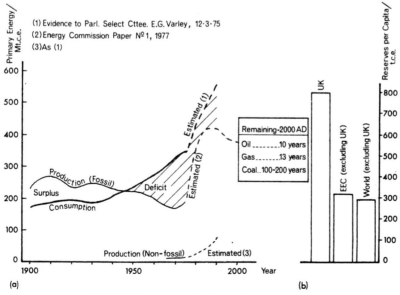

3.1 (a) UK energy production and consumption
(b) Fossil fuel reserves

Of the energy used in this country, 40% ends up intentionally or otherwise altering the air conditions within buildings. Clearly the body of professional men called the Chartered Institution of Building Services has a responsibility to the public to advise on its best use.

In the crisis conditions towards the end of 1973, a small group – the Emergency Committee of Eight – responded and issued a number of *Communiqués*. Later, the Technology Board constituted a formal Energy Executive Committee which sponsored a number of reports and what was later to become the CIBS *Building Energy Code*.

ENERGY POLICY

The philosophy behind the Code was engendered by the Institution's Energy Policy which may be summarised: conditions inside a building should be optimised for human comfort (or industrial process) with the minimal use of resources including energy; full advantage should be taken of all natural resources and phenomena; the form of energy to be used for any specific purpose in a building should be determined by the special suitability of that particular energy source; new buildings and their services should be so designed that they may, with existing buildings, be operated and maintained so as to avoid energy wastage; as much as possible of the energy used within a building should be recovered; the Institution would support any action by government to give positive encouragement to energy-saving measures; as appropriate the Institution would make known its views on matters affecting the proper use and conservation of energy in buildings; and, finally, the Institution believes that these objectives will only be achieved by co-operation of government, research organisations and all other bodies concerned with energy use.

SETTING PRIORITIES

The Institution recognised that its guidance would only have a short-term impact if it related to existing buildings of which about 80% will be standing in 2000 AD.

Early effort was therefore mainly directed towards the series of Energy Notes published in 1975-76. However, existing buildings may change their energy-using systems, so guidance on new installations also received attention. Note was taken of a draft circulated by the National Bureau of Standards early in 1974, which later became the well known ASHRAE Standard 90-75(1).

Recognising that building energy use is determined by the siting, shape and structure of the buildings themselves as well as the services within them and the use to which they are put, it was decided that likewise the CIBS Code must comprehensively consider the total building situation.

CO-OPERATION WITH GOVERNMENT

There are two approaches to conserving energy, the technical and the political. The former guides the designer towards more efficient and effective energy-using equipment and installations – the latter may, by controlling fuel supplies or specifying insulation standards, regulate the rate at which energy enters and leaves a building. Maximum benefit to the nation is likely to accrue from a combination of these two procedures. Regulations may take one of two forms, the prescriptive or the functional. As long ago as 1972, when the proposed Building Bill was circulated, the Institution commented that regulations should be in functional terms and the responsibility for compliance should be placed on the appropriate independent professional.

Much of this Building Bill has since been incorporated into the Health and Safety at Work etc Act 1974, which opened the way for regulations to be laid to conserve energy.

Any increase in prescriptive regulations brings problems of monitoring compliance. Recent proposals to amend Building Regulations make provision, for the first time, for a functional approach based on a performance specification – in other words, a form of building energy target.

The laying of such regulations would be greatly facilitated if a guidance code was already available and largely accepted by the professionals involved. The Building Regulations Division of the Department of the Environment had noted the CIBS code drafting activity and offered assistance towards reaching the goal more speedily.

OVERALL PLAN

The aim of the Code, agreed with DOE, is to set out comprehensive energy conservation guidelines to be applied to the design and operation of buildings and their services. Thus, from the outset, the Code is seen as applicable to both new and existing buildings. The Code is planned to be published in four parts, covering respectively — Design guidance, Design targets, Operational guidance and Operational targets.

Only Part 1 has yet been published.

Two difficulties presented themselves – first, the relative importance of the many factors which affect energy use for comfort in buildings; and the second that any proposed energy target must be achievable. Money was made available (to BSRIA) enabling the first assessment to be made and this information is being incorporated into Part 2 of the Code. Later, it is hoped that further money will allow collation and correlation of much of the factual data now beginning to accumulate on the energy use in existing buildings. This information would be the essential foundation for Part 4 of the Code. Although the Code excludes dwellings and buildings where there is a large proportion of process heat many of the recommendations will be found to be applicable.

It is no part of the Code to recommend a lowering of those standards which have come to be accepted as necessary for reasonable working conditions, neither is it part of the policy of the Institution to make recommendations which may save energy but in circumstances that involve a disproportionate cost in other resources, whether they be economic, social or environmental.

In the climate of the UK it is recognised that the general need is for buildings to be heated rather than cooled. It must also be recognised that refrigeration is relatively more expensive in energy terms. Within a building there are likely to be more wild, stray and casual energy sources contributing to heating rather than losses contributing to cooling. However, air conditioning can be essential in certain circumstances for instance, in urban areas where pollution and noise prevent windows being opened or in high rise constructions where windows have to be kept closed because of excessive wind speeds at the higher levels, or where a vital process contributes an excessive heat gain. The general tenure of the Code is that the justification for air conditioning should be technical necessity rather than a speculatively inspired financial exercise.

To summarise, the Code is designed to support the form of regulations now being proposed. Parts 1 and 3 of the Code detail mainly prescriptive advice for designers and operators respectively. Parts 2 and 4 of the Code will suggest target values of energy use for various classes of buildings again for designers and operators but which could be used as support for the functional type of regulation. It might be observed that the control of energy supplies to existing buildings by regulation is another name for rationing. Hopefully this will not be necessary and Part 4 of the CIBS Code will remain a useful tool for voluntary use.

TYPES OF ENERGY TARGET

It is important to agree exactly what is meant when the word 'target' is used since not only has the word itself been used to mean both energy used and energy saved but it has become confused with other terms such as budgets and audits.

Two types of target may be set, one relates to the call for energy which a system *could* make on the national resources under a certain closely prescribed set of conditions—this is termed a *demand target* in the Code—and the other relates to the amount of energy which a system actually consumes under the conditions of climate and use that prevail – in Code terms this is a *consumption target.* Within this definition, energy demand only can be estimated at the design stage and it can be considered a rate of energy use – ie power. Consumption on the other hand is energy per se. It must be realised that a building is unlikely to make the maximum demand on the energy supply represented by the summation of the total power ratings of all the equipment within it. Energy demands must therefore be estimated by considering a building's assumed operation over a period, conveniently a year. Energy demand targets are thus estimated averages and recognise a diversity of utilisation.

Buildings which exist, on the other hand can actually use energy and *consumption targets* apply to existing buildings.

PRIMARY AND SITE ENERGY

By no means all the energy stored chemically in fossil fuels is available for use in the building system, a varying proportion is lost in conversion and distribution.

In the case of building heating, about three times as much energy in the form of fossil fuels is required if the energy source at the building site is electricity than if it remains as fossil fuel. Energy in the form of fossil fuels is termed *primary energy*. Energy as electricity, coke, synthetic gas etc is *secondary energy*. Energy targets may be in terms of energy at site or as primary energy. Values of the ratio between primary energy and site energy (termed fuel factors) for various forms of site energy are listed in table 3.1.

Table 3.1   Values of primary energy per unit of site energy (fuel factors) (From BRE Digest 191)

Final energy form	Fuel factor
Coal	1.03
Natural gas	1.07
Oil	1.09
Other manufactured solid fuels	1.38
Town gas	1.42
Electricity	3.82

These are factual and determined from values of fuel use published annually in the *UK Digest of Energy Statistics*.

The objective of the Code is to save *primary* energy, and hopefully the Code will encourage designers to improve system efficiencies, exploit techniques of heat recovery and to further the use of non-depleting energy sources etc. It may be noted that since the unit cost of energy delivered to site at the moment roughly reflects the primary resource use, then a minimum energy *cost* solution is also likely to be minimum resource solution.

PART 1 – GUIDANCE TO DESIGNERS

This Part of the Code was published in November 1977.

In simple terms, in order to attain and maintain a comfortable environment inside the building, energy is added and distributed internally in a manner partly controlled and partly uncontrolled. (Uncontrolled energy arises from occupants, machinery, lighting etc.) External to the building energy is gained or lost in a largely uncontrolled manner (although solar radiation effects and air infiltration can be partly controlled) dependent on the vagaries of climate. Separating internal from external environment is the building structure, resisting the flow of heat but also storing heat within itself.

The structural characteristics are generally fixed at the design stage for the life of the building. The aim of comfort air treatment is to add or subtract in a controlled manner from the uncontrolled heat gains and losses so as to maintain constant internal conditions. The building is thus a dynamic whole and the Code breaks new ground by attempting to treat it as such. Very few will dispute this approach but it remains to be seen whether the joint design approach by architect, builder and engineer will be a commonplace occurrence.

Site and building

Often choices of sites present themselves and clients must be made aware of the benefits or penalties in long-term energy requirements which might arise from a particular choice. Appreciation of energy economics might lead in future to a change in urban skylines – some tower blocks tend to be more profligate in energy.

On building form, the Code draws attention to the energy balance between natural and artificial lighting as exemplified by narrow and deep plan building.

Glazing should not be dictated by appearance or fashion but be determined by technical considerations and made aesthetically acceptable by good design. Application of the Code should result in buildings having less than 30% glazing referred to the outside face and no glazing extending below the working plane level (fig 3.2).

The Code emphasises that lighting levels should depend on the task

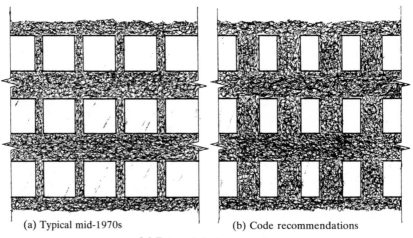

(a) Typical mid-1970s   (b) Code recommendations

3.2 External glazing amounts

3.3 Possible energy conserving building form (Courtesy Property Services Agency DOE)

performed and the levels recommended by the IES Code are seen as maxima. It is anticipated that the application of task lighting will increase and it is interesting to note that a Committee of the Institution is surveying the scope for development.

Another important innovation in Part 1 of the Code is the recognition given to the effect of the thermal response of a building. Thermal admittance as well as thermal transmittance is seen as being part of the designer's insulation considerations. The proposals in the extended Building Regulations for thermal insulation are seen to be minima and values approximately half of these are suggested in the Code as being attainable for the majority of heated buildings. These lower values are related to plant use.

Application of the Code could lead to building forms approaching that illustrated in fig 3.3.

Heating, ventilating and air-conditioning systems
Heating, ventilating and air-conditioning systems operate for most of the time at less than full load so designers will welcome the recommendation that performance data over a range of operating conditions should be furnished by equipment suppliers. But the incentive for it to be supplied will be in the hands of those same designers through their specifications. Load calculations are to be carried out as recommended in the CIBS Guide but Part 2 of the Code will take this further by specifying how calculations should be made to compare designs on the basis of assessed energy use.

The Code pays considerable attention to the control of systems and sets clear recommendations for heating and cooling controls. Although a large proportion of the thermal energy requirement of a building is devoted to warming or cooling outdoor air in very few installations today is this air treatment dependent on the building occupancy. The Code suggests that mechanical ventilation systems be so designed that treated air is shut off from unoccupied areas. This aspect of control alone could make a significant contribution to energy conservation.

The two most common media for transporting thermal energy round a building are air and water and their transport uses energy; the Code suggests limits to this, balancing the size of heat exchangers, pipes and ducts against fan and pump powers. The insulation of pipes and ducts is also given attention.

Guidance is given on boiler time-switching relating these times to the thermal inertia of buildings *and systems* as well as to the weather. Multiple boiler installations should be considered if this is likely to provide a lower seasonal energy use.

Basically, the Code recognises that to provide cooling is more costly in primary energy terms than to provide heating. Although a high co-efficient of performance can counteract the primary energy penalty of using electrical power it must be remembered that for most of the year the UK climate dictates heating. And if the required heat for comfort is more than adequately supplied by casual gains related to the use of the building then the first duty of the designer is to check if such gains can be reduced before installing cooling equipment. Sensible use of outdoor air for cooling may often provide a lower energy consuming solution.

From these considerations it is clear that no air-conditioning system should be designed to employ both heating and cooling at the same time.

Areas of buildings with different use and hence different heating/cooling load patterns should be considered and treated separately by the system designer. Many other detailed points are given but the underlying philosophy of *fitness for purpose* adopted throughout cannot be over-emphasised.

Minimum coefficients of performance are quoted in the Code and not surprisingly, bearing in mind equipment origin, these generally agree with those in ASHRAE 90-75. There is a small difference in presentation; ASHRAE 90-75 quotes two values – one for today's equipment and a higher value, for equipment manufactured from 1980 onwards. The CIBS Code recognises that larger sizes of equipment generally have higher coefficients of performance and it was felt that the ASHRAE 1980 values should be attainable by the larger equipment now. Except when passing through spaces to which they are supplying treatment air, leaking duct joints can be a source of energy waste and the Code recommends sealing low pressure duct joints. Additional sealing necessary to meet the requirements of this Code over and above that specified in Heating and Ventilating Contractor's Association Standard Specification DW141 may, of course, involve extra cost.

Hot water services
In the realm of hot water supply the Code again indicates areas for designer

choice, determined by overall energy use, whether or not to provide separate boiler plant, whether there should be central storage of hot water; whether the water circulation should be by pump and so on. The Code recommends that hot water should be stored at the temperature at which it is to be used rather than at the highest of the different use temperatures.

The Code recommends maximum lengths of pipe between point of use and point of storage, and maximum flow rates for shower fittings and hot and cold water taps, preference being indicated for the single outlet mixing variety.

Artificial lighting
The energy conscious system designer must be at home with the interplay between daylighting, artificial lighting and the thermal contribution to interior heat gains from the artificial lighting installation, so some lighting calculation procedures are included. Could it be that the amalgamation of IES into CIBS is not unrelated to considerations of conserving resources? The energy efficiencies of light sources useful in a commercial internal environment are still relatively low. All the energy supplied to the light source, including the luminous energy makes a contribution to internal heat gains.

The problem is to ensure that this heat is directed to the points where it can make its most effective contribution. Air heated by extraction through light fittings can often be re-used or some of the heat recovered from it by means of a heat exchanging device. The use of such 'air handling luminaires' was encouraged some years ago by a joint IES/IHVE Study Panel which published a performance test procedure for such fittings.

Other services
One of the first tasks of the Working Group charged with drafting the Code was to identify and rank in some order, the components of building services which contributed to energy use. Naturally, the bulk of the code is devoted to the major ones, heating, ventilating, air conditioning, hot water services, and artificial lighting. The final section of the Code makes reference to the relatively minor energy users – lifts and escalators, cooking equipment etc and the importance of power factor correction of electrical power supplies.

A summary of Part 1 of the Code would shout loud and clear the message that for optimum energy use no part of a building or its services, or the occupants or the occupation may be considered in isolation. As never before, energy conservation is a professional integrator.

The remainder of the code
The contents of that part of the Code already written and published have now been described mentioning some of the more important implications. Of the remainder of the Code, some is in draft, but some has not yet been considered.

What follows is then largely personal opinion but it may provoke some discussion and certainly comments made will be noted by the Working Parties and given consideration as work on the Code proceeds.

PART 2 – ENERGY USE CALCULATION AND DEMAND TARGETS

The objective is to suggest a means whereby designers of both buildings and services can, on the basis of its energy use effectiveness, compare one design with another or with a target energy demand derived by applying the proposed procedures to buildings designed according to Part 1 of the Code.

The procedure will include a method of computation and the enumeration of those factors of prescribed internal conditions, pattern of occupancy or process and of climate, over which the designer has no control. There thus remains scope for varying the building structure and the services installation within it so as to present the most energy effective system.

At the outset, the Working Group decided that there was great advantage in simplicity, and agreed to adopt conventional rather than computerised techniques. Differences in end result would be insignificant in the context of this part of the Code. The importance of distinguishing between estimates of energy use made at the design stage and actual energy consumption of existing buildings will be made clear. Part 2 of the Code concerns energy demand, already defined as the call for energy which a system *could* make on the national resources under a certain closely prescribed set of conditions. The calculation procedure is likely to break the system into its smaller component parts and for each determine a *calculated load* (L_c). For instance, the Code might recommend the calculation of design heat loss for a building (fabric and ventilation) to be according to section A5 of the CIBS Guide. Similarly for refrigeration, interior lighting, hot water service, lifts, pumps and fans.

Each of these loads would be adjusted by a *utilisation factor* (UF) to produce a *standardised load* (L_s).

$$L_s = L_c \times UF$$

Standard utilisation factors would be derived from estimates of hours of equivalent full load operation per annum, and would be tabulated. The Department of the Environment made available a sum of money to the Building Services Research and Information Association enabling the execution of a number of computer simulations specifically to estimate uncertain utilisation factors. Additionally, there are a number of other sources of such information.

Standardised loads, adjusted for the respective appliance efficiencies (η) result in *site energy demand* (SED).

$$SED = L_s \div \eta$$

Table 3.2   Possible arbitrary fuel factors

Form and use of site energy	Arbitrary fuel factor
Electricity – lighting load above 20 W/m^2	4.0
Electricity – lighting load up to 20 W/m^2	2.0
Electricity – space and water heating	4.0
Electricity – refrigeration	3.0
Electricity – pumps and fans	1.0
Energy from waste	0.5
Recovered energy	0.5
Non-depletive energy (on site)	0

The site energy demand relates to *primary energy demand* after adjustment with a *fuel factor*

Primary energy demand = Site energy demand × Fuel factor

Some typical fuel factors for different types of fuel are listed in table 3.1.

It is possible to allocate various values to fuel factors depending on the wish to encourage particular energy sources for particular uses. Thus, electricity for space-heating may be allotted a high fuel factor whereas electricity used for lighting, for which there is no other suitable alternative, could be allocated a lower factor. Similarly, energy recovered in a building which would otherwise be wasted may be allocated a fuel factor less than unity. Some suggested values are listed in table 3.2. These are necessarily tentative and values in Code Part 2 may well be different and indeed may be changed as time passes.

A summation of the various primary energy demands will give the *total primary energy demand* for the building. It is the intention to calculate total primary energy demands for a whole range of classes and sizes of buildings, these would become the primary energy demand *targets* against which designers could compare their particular designs. For ease of comparison such targets would be related to an appropriate building parameter (ie floor area).

It will be appreciated that a method of calculation such as that described depends upon the availability of acceptable utilisation factors, which in turn are dependent on the building shape, orientation and structure and the type of system installed. Clearly the Code cannot and need not define factors for every nuance of architectural whim, but is expected to cover cellular and open plan offices both low and high rise, single and multi-storey factories, shops and departmental stores of current general design, and several commonly encountered types of school and hospital. Sufficient utilisation factors will be tabulated for each of these to cover the majority of circumstances. Similarly, the range of targets postulated will be limited to the more common cases.

There will always be a need for buildings designed for circumstances not specifically covered by the general methods enumerated in the Code. The number of these is likely to be small so their energy drain on the nation's resources is not likely to be significant. Nevertheless, designers would be able to use much of the Code procedures and arrive at an optimum energy solution.

PART 3 – GUIDANCE FOR BUILDING OPERATORS

The aim is to provide for owners and operators of buildings authoritative guidance that hopefully has already been provided for designers in Part 1.

The Working Group are preparing recommendations covering energy conserving management applicable to most types of buildings. The proposals will be presented in a manner which can be easily understood by building operators not specifically trained in building services design.

Measures for energy conservation are categorised as short, medium or long-term – the time-scales being generally related to the time to put into effect and the pay-back period. Suggested periods would be a few months, several years, and over five years respectively. Long-term measures are dependent on the development of special techniques and equipment (such as certain heat recovery devices) and are not considered in the Code. However, such aspects are not being neglected by CIBS and other groups undertaking studies and are expected to report within the next year or so. The Code then, considers short and medium-term measures, as did the earlier Energy Notes. (2).

The recommendations listed as short-term generally cost very little and may be put into effect without detailed financial evaluation. The medium-term actions should be preceded by a detailed assessment since it is of little use implementing a proposal if it results in a disproportionate cost in other resources.

Energy conservation is an on-going activity and savings from one measure may advantageously be invested in the next. In this way capital outlay is kept to a minimum yet increasing benefits may be reaped.

As with the earlier parts of the Code, attention is given to the four main influences which control energy consumed by a building:
- The climate outside the building
- The building structure
- The efficiency and effectiveness of the plant and equipment generating and distributing the thermal energy within the building
- The pattern and duration of use within the building.

The Code considers the scope available to the building operator to modify these parameters. This ranges from nothing but an understanding of the effects in the case of climate, to total control (shut-down) of building use.

The Code will not ignore the management aspects of instituting an energy conservation programme and will suggest the necessary steps in undertaking surveys and will contain examples of the types of report documentation considered essential to the systematic control of energy use in buildings.

PART 4 – COMPARING ENERGY USE AND CONSUMPTION TARGETS

The energy actually used is often known, if only from a summation of the bills received from fuel suppliers. Because of the day to day variations in climate and use, it is necessary to adjust 'as read' energy data before valid comparisons can be made. Comparisons of energy use may show whether any significant change has occurred in the plant operation and function. Again, the energy used by one particular building during a certain period might be compared with a target value specified for the general type of building and use pattern under a standard set of climatic conditions, thus highlighting shortcomings in the building or system design or its operation as evidenced by the standard of 'housekeeping'.

Part 4 of the Code will follow the pattern established in Part 2 and outline relatively simple computation procedures for energy comparisons. Hopefully, it will propose targets based upon factual measurement and not upon idealised simulations. In this connection, Building Services Research and Information Association have already undertaken a feasibility study and have confirmed that in recent years and in various places a very considerable amount of energy use data has been recorded for a wide range of buildings. This data would need to be standardised and correlated with chosen building parameters for the various types of building. It is hoped that the Department of the Environment will make money available for this purpose. However such a project is likely to take up to two years and Part 4 of the CIBS Energy Code is unlikely to be published before early 1980.

THE WAY AHEAD

This chapter is a brief description of the philosophy underlying the CIBS Energy Code and shows how its publication is being achieved step by step. Although, during the last few years the amount written and advice given on matters related to energy use has reached prodigious proportions, no other Institution has attempted to systematically assemble authoritative recommendations of proven effectiveness for the benefit of its Members and the public at large. Only in the US has ASHRAE attempted a similar task.

Like CIBS they have already issued guidance to designers and have a number of Sectional Codes in draft form covering guidance to operators of existing buildings. As far as it is known, ASHRAE has not yet agreed standardised calculation and comparison procedures and resulting energy targets.

This country is better placed than most with its endowment of energy resources but never again will fossil fuels be plentiful: within a decade restraints will be felt. Initially the deterrent to wasteful energy use will be financial but if that brake is insufficient then the deterrent may have to become physical rather than fiscal.

In its Energy Code the Institution has tried to cater for both these eventualities – on the one hand guidance is being provided to assist in avoiding the worst effects of increased energy cost and on the other hand if energy use has to be restricted then realistic targets are being proposed as bench-marks for control.

Undoubtedly the richest source of energy for the '70s and '80s is conservation.

ACKNOWLEDGEMENTS

The author recognises that without the work of CIBS Members serving on the various Code Task Groups this chapter could not have been written and he acknowledges the very helpful comments received during its preparation.

REFERENCES

1 The American Society of Heating, Refrigeration and Air-Conditioning Engineers Inc, *Energy Conservation in New Building Design – ASHRAE Standard 90-75.*
2 Chartered Institution of Building Services, *CIBS Energy Notes* : Offices; Factories; Schools; Shops; Sports Centres.

DISCUSSION

W. R. H. Orchard (Orchard Partners)   Since the publication of Energy Paper 20 in 1977 which established the long term economic case for combined heat and power, there have been two significant changes: 1. The discount rate for nationalised industries has been reduced from 10% test discount rate to 5% real rate of return, and 2. the Department of the Environment now agrees that fuels will double in price in real terms by the end of the century as opposed to staying level. As a result, both small and large cities are economic and suitable for combined heat and power and I expect that shortly a recommendation will be made to Government to implement combined heat and power nationally.

Mr Peach has produced ratios for different energy sources and I agree that the ratios he has for reject or waste heat would apply to combined heat and power as there is an 8:1 ratio between the long term marginal cost of heat from a low temperature combined heat and power plant and electric power at full load.

The histogram fig 3.4 indicates annual marginal costs of heat produced from fuels by other methods and the relevant ratios. I suspect that combined heat and power alters some of the assumptions in the CIBS Code and I would ask how the CIBS proposes to modify its Code to take account of combined heat and power and the effect combined heat and power will have on the question of optimisation of air conditioning and heating to save energy.

J. Peach (The Chartered Institution of Building Services) I think this is one example of how basing an energy fuel option on a relatively simple factor can be useful if everyone can agree such factors for the various fuel options. It does have the advantage of speedy computation compared with cost comparisons, from the point of view purely of optimising a design for minimum energy use. But exactly what the factors would be in the case of combined heat and power I could not say, although there is some indication on your illustration. This is something which would have to be considered when the time comes (of more general use of combined heat and power), which the Institution, I think, would welcome sooner rather than later. You will know of the first interim report on combined heat and power.* This contains, as an appendix, a contribution made by the Institution.

I am afraid I cannot say exactly how an allowance will be made in the calculation procedure because that calculation procedure is not yet finalised for the cases of the more normal energy sources.

W. R. H. Orchard I suggest two codes wil be needed, one for combined heat and power areas and another for areas where combined heat and power will not exist such as rural areas. Combined heat and power in cities linked to electric night storage units or electric heat pumps in rural areas will almost certainly provide an overall optimum use of energy resources in which case, for electricity there should be a lower penalty in the code for rural areas than for cities.

J. Peach Air conditioning is the subject of this discussion. Consider an air-conditioned building where the energy source is hot water from a combined heat and power system. Would Mr Orchard suggest that the source of energy for refrigeration should also be hot water – say by using absorption refrigeration equipment? Depending on relative fuel factors this might be the correct option.

W. R. H. Orchard Electric air conditioning in combined heat and power cities may be economic in energy terms if the electricity powering the air conditioning matches the city's demand for heat for domestic hot water. Alternately, absorption air conditioning operating from the district heating network may be even more economic in overall energy terms if the higher grade electric energy can be used for other purposes. The CIBS Code assumes that in the case of electrical energy the remaining energy has been lost. This does not apply with combined heat and power.

To indicate the kind of change that combined heat and power can make to calculations the marginal cost of heat for combined heat and power is a third of that for a district heating boiler. This will change the economic thickness of insulation of a building for the two alternative sources of heat.

E. C. Lovelock (Shell UK (Admin. Services)) To illustrate this point, would it be true to say that if a district heating system were utilising waste heat it would be more appropriate, in the cause of energy conservation, to

* Energy Paper No 20, appendix 9 Department of Energy Publication *District Heating Combined with Electricity Generation in the United Kingdom.*

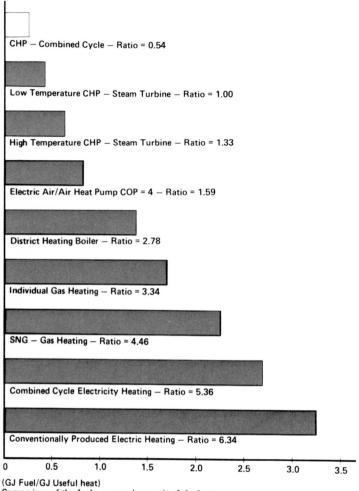

CHP — Combined Cycle — Ratio = 0.54

Low Temperature CHP — Steam Turbine — Ratio = 1.00

High Temperature CHP — Steam Turbine — Ratio = 1.33

Electric Air/Air Heat Pump COP = 4 — Ratio = 1.59

District Heating Boiler — Ratio = 2.78

Individual Gas Heating — Ratio = 3.34

SNG — Gas Heating — Ratio = 4.46

Combined Cycle Electricity Heating — Ratio = 5.36

Conventionally Produced Electric Heating — Ratio = 6.34

0   0.5   1.0   1.5   2.0   2.5   3.0   3.5

(GJ Fuel/GJ Useful heat)
Comparison of the fuel consumed per unit of the heat
produced for various methods of heat production.

3.4 Comparison of fuel cost component of the total unit heat production cost for various heat production methods (expressed as a fraction of the unit input fuel cost)

Assumptions for preparation of histogram showing the long-term marginal cost of heat for alternative systems

(All figures based on gross calorific value of fuel)

1. No capital or pumping costs for gas, district heating and individual heating systems are included.
2. Transmission losses are not included for natural gas but are included for synthetic natural gas.
3. Annual efficiency of electrical generation   35%
   Annual efficiency of district heating boiler   80%
   Annual efficiency of individual gas boiler   60%
   Distribution loss for district heating network   10%
   Distribution loss for electrical network   10%
   Distribution loss for SNG grid   6.7%
4. Low temperature district heating from a combined heat and power back pressure turbine.

use absorption type refrigeration rather than electrically driven compressors, this being the best solution in terms of the total energy system?

W. R. H. Orchard Precisely, and that might well determine the temperature of this distribution system.

In Canada, the development of absorption systems operating on lower grade heat is actively being investigated. The use of a lower primary temperature, say around 95°C, would enable electricity to be generated at higher efficiencies. This can provide overall, more attractive energy use with, for instance, absorption air conditioning plants being used in cities from the reject heat and electrical air conditioning units used in rural areas.

J. Bonthron (University of Strathclyde) Mr Peach said that a future code would provide a method of comparing systems and that this will be a manual rather than a computer based technique. Messrs Smith and Hammond (see Chapter 5) could not possibly have given the detailed comparison between systems without a computer which is a valuable tool for energy optimisation. The International Energy Agency is investigating computer based system comparison techniques including an exercise with the ABACUS Unit at Strathclyde. Is the manual method an initial approach which can be supplemented by a computer based method in the future?

J. Peach The decision when we embarked on the Code was to concentrate initially on a manual method, on the basis that it could be applied more readily by a larger number of people and would probably be quite satisfactory, particularly for the large majority of buildings with heating alone. For the more sophisticated buildings, a computerised procedure might in the end be desirable. It could be that there might be an appendix to provide a bridge between a manual and a fully computerised technique. Initially, though, we are concentrating on a manual method where the utilisation factors have in some cases been determined from computer simulation.

J. M. Cooling (Balfour Kilpatrick Ltd) Mr Peach told us of the relationship between the CIBS and the Department of the Environment and how they

---

Water temperatures 90°C supply, 50°C return. Electrical efficiency is 26.7%. Overall efficiency of combined heat and power production 80%.

5. High temperature district heating from a combined heat and power back pressure turbine. Water temperatures 120°C supply, 70°C return. Electrical efficiency is 23.1%. Overall efficiency of combined heat and power production 80%.

6. The SNG production plant is assumed to be 80% efficient with a 6.7% distribution loss and 60% average annual coversion to heat at the local domestic boiler.

7. The calculation of the marginal cost of heat from a combined heat and power station is based on the fuel consumption at the CHP plant to produce one unit of heat at the consumer's premises, minus the energy that would be consumed to produce the electricity from the CHP plant at 35% efficiency.

8. The annual efficiency of combined cycle electricity generation is 41.5%.

9. The marginal fuel consumption from a CHP combined cycle plant was calculated on the basis that the heat production requires only increased supplementary firing of the heat recovery boiler. No credit has been taken for the higher efficiency of combined cycle generation over more conventional means. If this credit were taken into account, the relative marginal cost would be -0.33. (This figure was supplied by W. R. H. Orchard.)

have worked together on the Code. I would like to explain how the CIBS is involved in consultation with the Department of Energy. The Secretary of State for Energy has an Advisory Council on Energy Conservation and CIBS are represented there, not as a body, but by me in my personal capacity. The Advisory Council has four working groups. One of them is the Working Group on Buildings which is chaired by Professor Pat O'Sullivan and its report to the Minister has been published recently. The working group on buildings has set up four working parties dealing with various aspects of energy conservation in buildings. One looking at structure is, chaired by Professor O'Sullivan. Another, which I chair, deals with systems and controls and on it are represented most bodies who are interested and would wish to be consulted. Jack Peach represents the CIBS on that working party, Peter Coles represents the Building Services Research and Information Association, the Watt Committe, HEVAC, HVCA and ECA and the Plumbers' Federation are all represented too. I intend in the future to consult with interested parties on any measures which the Minister might bring in.

Returning to the Code and the introduction of energy targets which will be contained in Parts 2 and 4 of the Code, the CIBS intend that the targets in the Code will be advisory and for use in comparisons. We should bear in mind that there could come a time, if the Government believe this is a good way of conserving energy, when such targets might be made mandatory by introducing appropriate legislation. Certainly the design targets which will be a reflection of building regulations are in a different form virtually mandatory now. In setting operational targets the CIBS must consider very carefully how these targets would be implemented if they became mandatory.

J. Peach   It is important to distinguish the targets which will be in Part 2 of the Code for design purposes and those in Part 4 which will relate to existing buildings. The design targets – or demand targets – are based on a range of standardised classes of buildings with the targets derived by application of the CIBS method of calculation to buildings which comply with Part 1 of the Code. They are intended purely for purposes of design comparison and this must be made clear. Obviously, it is hoped the figures obtained in this way will not be too far out.

W. P. Jones (Haden Young Ltd)   As chairman of the working party producing Part 2 of the Energy Code I would like to comment on two points that have been raised. The first was on the question of modifications to the Code and the possible publication of an amended version to take account of combined heat and power techniques. I want to emphasise that the working party is conscious of the fact that the Code cannot be thoughtlessly applied to all parts of the country and to different systems, without taking account of the technical variations in design and without taking account of the different environmental conditions that may exist, in different parts of the country. We are therefore proposing methods of relaxing targets to take account of urban pollution by noise and dirt which could mean that windows would have to be kept closed, and even double glazed. Consequently a building in such a dirty, noisy, urban area would have a little more freedom for the calculation of a demand target than would say, a low rise building in a country area. Equally we would try and take account of the fact that combined heat and power generation is a more attractive and energy useful technique than other methods. Mr Orchard can rest assured this has not escaped our attention.

The second point was on computerised methods. I must say that I am a great believer in simple methods of calculation. Having considered over the last few years many programmes put forward to assess energy consumptions for buildings and systems I have found it difficult, when presented with a programme, as a *fait accompli,* to weigh up its merits. There is a great virtue in simplicity and because we want the Code to be used I believe that proposing simple methods will be the best stimulus. I am therefore a strong supporter of the manual method but that is not to decry computerised calculation. We recognise different systems will have different energy consumptions which, if worked out to the nth degree, would indicate different demand targets in watts per square metre of floor area. In the early work on Part 2 of the Code use was made of a study by Building Research and Information Association on behalf of the Department of the Environment* on the relative energy consumption of different systems with particular reference to the utilisation of refrigeration plant: it showed surprisingly small variation. The ancillary items of equipment such as fans (particularly), pumps and electric lighting, account for a very large proportion of the energy consumed. Because these are fairly constant loads and represent a weighty factor in estimating the demand target I think it is reasonable to argue that a computerised version for calculation is not necessary. This supports the view that a simplified version of calculation would help the Code to be used.

J. W. Coe (Standard and Pochin Ltd) As manufacturers of air-conditioning equipment we deal on a day to day basis with a broad spectrum of the trade, mostly with non-technical staff. What our enquirers do is generally dictated by instruction, commercial necessity or clear specification and in most instances we are not even able to have a conversation about higher efficiency selections resulting in lower running costs. The interest is invariably in capital cost alone.

Because enquirers provide inadequate information the manufacturer is forced to carry a large technical responsibility in respect of central plant which accounts for a major proportion of electrical costs. Unless there are clearly defined requirements and efficiency there are a wide range of options for a given output. Do we choose low efficiency to be more competitive on prime cost or choose best efficiency because of national interest and lose the order?

*Not published.

How can implementation of the CIBS Energy Code be obtained without legislation? How can both building owner and designer be rewarded for saving energy by following these important new codes?

J. Peach Concerning inducing people to use the Code; we did, at one stage, consider how it might be possible to introduce cost factors. This was because making a comparison solely on a primary energy basis, as we are deciding to do, does not allow a comparison to be made relating say, the maintenance cost of a piece of equipment with its energy cost, during its lifetime. One can have a conflict between optimising energy use and optimising maintenance. We cannot consider that option in the method we are proposing.

The main options in design are the kind of system, the fuel and the type of structure of the building (assuming climate and use to be constant factors). The method we are proposing does allow those to be inter-related on the basis of energy use but not necessarily cost.

E. C. Lovelock The justification for higher investment in the cause of energy saving can only be on the basis of an acceptable pay-off, taking into consideration any tax advantages or legislation which the Government may introduce. I think most Companies are looking for a pay-off of something like three to five years for increased expenditure for energy saving purposes. I would have thought that if you could demonstrate that by a slightly higher capital cost that incremental cost has a pay off within three to five years then the proposal would gain acceptance. To take electric lighting as an example there has been no difficulty in selling new installations using fluorescent tubes to replace incandescent lighting because the saving is easily demonstrated. Now the second generation of fluorescent tubes are coming in which are more efficient and although their initial cost is higher than the first generation tubes they find a ready market because of their low energy consumption and therefore saving in this respect.

In a highly competitive market however, one can see that the seller must be in a dilemma as regards the basis for making up a tender. Possibly it means putting in both the lowest figure, disregarding energy consumption and then the higher figure having regard to energy conservation showing the justification for the higher spending.

J. W. Coe Mr Peach has indicated that 80% of existing buildings will be standing in 2000 AD. Probably the majority of these buildings will have services provided without engaging Consulting Engineers and the standards of the installation will be as widely varied as the quality of those carrying out the work. What inducement is there to pay regard to higher efficiency selection or to features which would help maintenance?

Our experience is that the usual enquiry requirements are quick response with lowest price and best delivery. It is only on Authority or Consultant prepared work that equipment efficiency is stated, or implied by restriction on air and water velocities or that features are stated in detail. It is rare indeed for a user orientated specification to reach the equipment manufacturer unless there is someone in between establishing standards. Education in this respect is therefore very important and those responsible for the eventual running costs need to be encouraged to ask pertinent questions in the first instance.

4 Designing buildings for air conditioning

DAVID ALLFORD

INTRODUCTION

Architects today have to solve problems of increasing complexity. Clients have become more demanding in their requirements as their knowledge of their own needs has improved. Governments are increasingly diligent in seeking compliance with ever more sophisticated health and safety regulations. Innovations increase the range of choice for the solution of technical problems, whilst the additional need to take into account the energy implications of design might reasonably have been expected to throw the architect off balance.

Much of the current building research effort into energy conscious buildings has been initiated by architects. Whether the imaginative studies of wind and solar power and of autonomous dwellings will have any lasting effect upon building technology must await a full assessment of their long term economic worth. Few clients, including Government, can make decisions about investment in capital stock that do not reflect current market prices and no reliable way has been found to introduce the cost of resource depletion into investment decisions, despite the widespread acceptance of the principle of husbanding our natural resources.

Building design impinges upon resource use in many ways. Energy is required both to make the components of modern construction and to provide air, light and other essential mechanical services for the building in use. It has been estimated that for a domestic house the capital energy cost is exceeded by the running cost after only five years. If heating is provided by electricity then only 18 months are needed for running energy costs to equal capital energy costs. Much remains to be done in the accurate assessment of energy inputs into building materials and although running costs are confirmed as more important in energy terms, the development of reliable guidance on energy intensive materials is long overdue (1).

It is difficult to assess the effect of design decisions made now upon the rate of resource depletion in the future. Real effects will be influenced by political and economic factors and by the search for alternatives to non-renewable resources. The uncertainty about such factors makes the architect's task especially challenging as it requires a recognition that the future may make unpredictable demands upon buildings. Some uncertainty is reduced by the advance of scientific understanding, but our use of this convergent knowledge must be tempered by a more divergent awareness of all that the future may have in store. In practical terms, is it possible to design buildings now that will accommodate unforeseen needs and that can be environmentally tuned to accord with these needs?

As understanding of individual facets of building performance increases in sophistication and precision, there has not been a concomitant increase in understanding their interdependence, or in understanding the importance of the uncertain conditions in the future which will affect the use and usefulness of a building. Two examples of direct relevance to this paper are the unknown future prices and availability of fuels and the changing expectation people may have for the quality of their environment.

The kinds of knowledge that affect architecture are many and varied. Architecture is not an exact science and lacks a comprehensive tradition of theory. It has drawn upon physics and chemistry and the engineering sciences to secure safe and stable structures, upon the social and environmental sciences to secure reasonable comfort and upon operational research to achieve efficient spatial organisation. As knowledge increases the social and economic climate also changes. These changes cannot be expected to occur in any ordered sequence. Yet the architect, and the whole design team, must always be able to absorb new intelligence, weigh its relevance with what is already known and act decisively while acknowledging the areas of uncertainty that remain uncharted.

COPING WITH THE FUTURE

In dealing with the analysis of client's requirements few architects today would fail to raise questions about the client's future plans. By examining the processes of organisational growth and change, the interaction of an organisation with its environment, it has been possible at least to grapple with anticipated change and to adopt reasonable design strategies to cope with the unanticipated. By designing spaces capable of many uses, by allowing for internal change and by allowing where appropriate for contiguous extension, architects have been able to develop reasonably robust designs which should be able to respond to changing needs (2). Engineering plant in buildings has a much shorter amortisation period than the buildings themselves, and it might be argued that it does not need to be designed to cope with change if it is to be replaced or modified fairly frequently.

Obviously the most interesting question that arises here is whether the decision to have an air-conditioned building can itself be robust. After all, there are few circumstances in which air conditioning would be chosen at all if energy conservation were the single most important criterion. Perhaps, considering possible fluctuations in the price of basic fuels, we should be designing buildings in such a way that we can use the relatively cheap fuel now to provide high comfort conditions with air conditioning whilst making it possible to remove it later if price rises logically so dictate. After all, working in developing countries it is standard practice to enhance standards as resources become available. It is surely not illogical to make provision to reduce standards as resources become scarcer.

ORIENTATION AND BUILDING FORM

Orientation

The belief that air conditioning obviates the need to consider orientation and liberates the plan from the narrow limits of what could be naturally lit and ventilated is clearly fallacious. Orientation can affect the way in which the sun influences internal comfort conditions. In cold climates the maximum solar radiation will be sought and in hot climates the minimum. Not only orientation will be used to achieve these objectives. Dark colour treatment may enhance the effect of the sun, and many devices are used to help reduce it as appropriate. The impact of orientation and building form upon the efficiency with which a building can be air conditioned is considerable.

While recognising this the architect also has to consider orientation within the practical limitations imposed by the site. In an urban area for example, his choices may be severely limited by building regulations, planning controls, access, views and existing environmental hazards like noise. It may be possible to exploit the density of the surrounding area by using existing buildings to provide shading. It is unlikely in these circumstances that orientation will be solely determined by climatic factors. In rural areas, and especially in developing countries where land is reasonably available and site constraints less severe, climate alone may well be the dominant factor in orientation. Exposing a building to the sun unnecessarily may create problems that no combination of shading devices and mechanical plant can readily be expected to solve.

Further, in complex buildings like hospitals not all accommodation may be air conditioned, and orientation will be chosen primarily to minimise solar radiation on the faces of naturally ventilated spaces and, of course, to gain the best natural movement from the prevailing pattern of wind. This will incidentally benefit the air-conditioned buildings as well, if they are similarly orientated.

Building form

Some functional needs largely dictate the form of the building which accommodates them. Others may be less demanding. The intricate procedures of the diagnostic and treatment departments in the modern hospital or the production process in a modern factory will have an overriding impact upon the eventual form of their buildings, while office accommodation may impose fewer constraints on building shape.

The space needed for air-conditioning plant itself may influence the shape, and indeed the size, of buildings. Plant must be accommodated and air must be delivered to and retrieved from individual rooms.

While taken for granted in theory, the satisfactory integration of structure and services is rarely achieved in practice. Solutions range from simple combinations of vertical and horizontal spaces for air supplied from a plant room to the provision of complete interstitial service floors serving every level. The choice will depend upon the complexity of servicing envisaged, the likelihood of changes being needed, capital and running cost implications, physical constraints like height restrictions, and whether the activities being served can be interrupted for servicing. Air-conditioned buildings need more plant and service zone space than non-air-conditioned buildings and the plant gives rise to acoustic problems that must be taken into account. The integration of heat rejection equipment into the structure, which can affect the appearance of a building, poses particular problems. Cooling towers can often disfigure an otherwise ordered appearance and, in addition, discharge visible plumes of moisture-laden air. Air-cooled condensers, requiring more space and making more noise than cooling towers, raise even more difficulties, that must be anticipated at the design stage.

As with orientation, considerations of siting and density will have to be taken into account and the impact of the building upon its surroundings, including the micro-climatological effect, properly assessed. In some situations the architect may have little choice open to him in the face of urban design criteria established by planning authorities, although there are welcome signs of greater flexibility in the official interpretation of such criteria.

Setting aside the obvious practicalities of building form, theoretical analysis has been undertaken of the way in which building shape may reduce heat losses. Heat loss is dependent upon the surface areas and their thermal transmission characteristics. Assuming a rectangular shape and equal thermal transmittance for walls and roof, with zero transmittance through the ground, March has shown that the best shape is square on a plan with a height equal to half the length of a side – a half cube (3). But, as March says, 'different walls will have different average transmittance values not only because they may be constructed differently with varying proportions of fenestration but also because of their exposure and orientation which gives rise to particular external surface transmittance values' (4). When this is considered, he concludes that 'if opposite pairs of faces have a different mean transmittance value the form which will minimise heat losses . . . is the built form whose thermal image is a cube such that heat losses through all three pairs of opposite sides are equal' (5). This kind of analysis is most stimulating in countering the more easily understood rules of thumb that a generation of architects and engineers have come to take for granted.

Future change

Can decisions about orientation and built form leave any scope for future change in response to changes in energy availability or cost?

At first sight the answer appears to be 'no'. Buildings generally stay where they are put. There are exceptions of course. The summer house on wheels that is able to follow the progress of the sun is an obvious example, as are designs for caravans, from Buckminster Fuller's Dymaxion houses (dating from 1928, 1940 and 1944) to the modern air-conditioned mobile home (6).

But buildings do not necessarily stay the same size or shape. Many studies have shown how buildings grow in response to organisational needs; new accommodation is added as an extension to existing buildings and even new floors are built on top.

While taking these possible growth expectations into account from the start, the architect helps the client cope with his future functional needs, it is difficult to say if this can help him save energy, other than by enabling his new buildings – built presumably to more exacting energy requirements – to work in association with his old.

BUILDING ELEMENTS

Thermal transmission

Having considered thermal transmission in terms of building form it must now be considered as an important factor in detailed fabric design. As Knight has observed 'people have in the past instinctively built their houses and offices to standards of thermal insulation in almost direct proportion to the severity of the climate' (7). The proportion of total energy produced which is used to heat buildings, about 40–50%, is the same throughout East and West Europe and not much different in the colder areas of USA and Canada. Different energy costs appear to have played a minor part in determining the amount of energy spent on space heating, and until recently have had a modest influence on energy conservation policies.

The current practice of assessing the heat transfer through the fabric in order to establish building materials and engineering specifications is normally undertaken by steady-state calculation which assumes constant heat transfer through the fabric and ignores fluctuations of environmental influence. Past experience suggests that air-conditioning systems designed by this approach using fixed winter and summer design temperatures are liable to be oversized and unable to provide optimum conditions.

In a YRM Study for the Department of Health and Social Security new techniques which seek to overcome the limitations of the steady-state approach were used (8). They have been used on most projects undertaken by YRM in warm climates. The techniques rely upon a computer programme and a memory bank containing data on the dynamic thermal performance of the building fabric affected by diurnal and seasonal fluctuation, surface absorptivity of the fabric, input through lighting and human occupancy, percentage of glazed area and solar shading devices (9).

Applying the technique to the hot and arid climate of Saudi Arabia it was found from the computer simulation that a well balanced building could almost provide an internal environment within the human comfort zone without any mechanical means of cooling. In proceeding to establish the energy requirements, taking into account the dynamic thermal performance of the building, it was found that a 30% reduction in energy needs was possible when compared with the load calculated by the steady-state approach. Substantial savings in capital and running costs of air conditioning were estimated.

The optimum behaviour of building fabric depends upon a careful balance between the thermal resistance provided by insulation and the thermal capacity provided by mass in conjunction with the absorptivity of the fabric surface characteristics. Each climatic zone, even each microclimate, requires a particular balance between resistance and capacity. Further, the effectiveness of a composite element of the building fabric in reducing internal surface temperatures depends upon the relative location of the insulation to the mass. The thermal storage effect of a building depends primarily on the mass of the structure rather than the insulation provisions. Mass can be built into the external structure to defer the peak heat gain penetration to a time suitable to the occupancy of the building. This may also be employed to reduce the requirements for cooling.

Fenestration
The problem of window design has always excited architects. Windows establish a relationship between the interior of the building and its surroundings by providing views and visual variety: they provide daylight, a sense of time and a sense of season. They also admit solar radiation and allow heat to escape. They have higher thermal transmission values than are attainable with opaque walls and are the most obvious target for those single mindedly pursuing the reduction of energy input for buildings.

As Hardy and O'Sullivan argue 'reducing the area of glazing has a marked influence . . . as not only are the heat gains reduced, but also the heat losses, and the replacement of glass by wall material increases the overall thermal capacity of the external wall' (10). This is followed in their influential study by some extraordinarily subjective views in support of narrow vertical windows, the lack of supporting evidence for which does much to undermine confidence in their general thesis. Psychophysicists who take leaps in the dark towards design solutions do so not only at their own risk, but at the risk of building developers and users who may suffer the results of a simplistic reading of such studies. Coupling reduced window areas with permanent artificial lighting, Hardy and O'Sullivan conclude that there would be no space heating requirements and a low continuous cooling load. 'As the energy requirements for such a building have been shown to be less than that required for a building designed for daylight, there appears to be an economic case for studying the design of lighting installations in relation to the heat output and the avoidance of additional space heating' (11). Several factors will illustrate how important it is for designers to understand fully the context of assumptions underlying such studies:

(i)   Conclusions are drawn for broad application on the basis of measurements taken in three rooms only.

(ii) The views expressed about window desirability are not substantiated statistically, and barely in an anecdotal way, by the very small sample of user reactions elicited.

(iii) The degree to which some important variables are subject to change over time is overlooked. Most obvious is the varying real and relative cost of fuel, labour and building materials, which would, if taken into account, have an enormous effect upon conclusions that could properly be drawn from psychophysical investigations. Again, to assume that the thermal transmission characteristics of glass are going to stay the same is simply absurd. New types of double glazing are expected to become available in the not too distant future that may reduce transmittance from the current 2.8 W/m² °C, to about 1.0 W/m² °C – equalling, that is, the present performance of cavity walls built to Building Regulation Standards (12). Glazing units which will be twice as efficient as presently available will change the design of the external fabric drastically, especially in air-conditioned buildings. However, the choice of single or double glazing for air-conditioned buildings is often made upon purely economic criteria – indeed in some cases apparently irrational decisions are made for fiscal reasons by clients operating within particular taxation constraints. It is difficult to prove a purely economic case for selecting double glazing with the present fuel tariff operating in the UK. It is understood that the Department of the Environment in conjunction with the Building Regulations Advisory Board are preparing new mandatory stipulations for buildings and this may lead to more regular use of double glazing in the future.

(iv) Important design considerations are simply ignored or gratuitously over simplified. For example, in the legitimate pursuit of reasonable comfort conditions, more elusive but perhaps more important features may be overlooked. As Schneider has recently said 'Health and comfort are often contradictory. Whatever we call comfort these days may often result in sickness. Take for instance the static unchanging climate, devoid of stimulus, of an air-conditioning system in which temperature and humidity are being kept constant' (13). Other critics have noted the need for stimulus and choice in the urban environment (14). In this context, the design of windows takes on a significance far wider even than the reduction of energy inputs. It establishes the overall visual context in which human activities occur, guiding unconsciously important patterns of behaviour and sensation.

Shading

The sun's spectral composition comprises three ranges of waves and each affects buildings and their occupants in different ways. The shortest wave in the ultra-violet region has very important therapeutic value but is filtered out by ordinary glass. Most shading devices re-radiate ultra-violet waves. The visible range of the spectrum is the middle range. Shading devices have to ensure the admittance of sufficient illumination to a building while reducing glare to a tolerable minimum. The infra-red range is responsible for the heat impact on buildings and from the point of view of environmental comfort has the strongest impact upon air conditioning.

Shading devices must be designed to allow as much sunshine to reach the building as possible during the cold season while ensuring maximum re-radiation and prevention of heat penetration during the hot season. Successful designs which meet these apparently contradictory criteria rely upon analysis of the changing azimuth and altitude of the sun during the year. A range of both vertical and horizontal shading devices working with the 'automatic' seasonal control will need to be established for each elevation of the building. There are no comprehensive solutions that perform in all climates.

There are various approaches to establishing the best form of shading. Direct observation is time consuming and probably unreliable. Computer techniques employing graphic input for the evaluation of fixed multi-fin shading devices and shadow displays have been used successfully. Manual graphic presentations, using shadow angle protractors as overlays to solar charts, allow both vertical and horizontal shadows to be constructed.

Seldom satisfactory in a warm climate is double glazing with an interspace blind. Double glazing leads in most instances to an increased greenhouse effect. Two-thirds of the heat transfer between double glazing derives from radiation, some portion of it being re-radiated through the use of a blind. But one third of the heat transfer is caused by convection and that transfer will be unimpaired by the use of an internal blind. External shading devices are, therefore, preferred. Solar reflecting glass is an ideal substitute where solar shading is not possible but even the most efficient reflecting glass will not produce the same effect on reduction of direct radiation as external shading.

Future change

There are many ways in which building elements may need to respond to changing energy conditions in the future. Thermal transmission, fenestration and shading need to be considered. As Fisk argues, 'the cost of keeping a future option open may even be incurred as a real material cost associated with design. In terms of constructing a building for the energy future, design decisions are not always as stark as sometimes thought. The design decision is not, for example, whether or whether not to install a solar collector, but how much to spend on keeping the option open of installing a collector. It may be the best way to do this is to install a complete collector system, but this may not necessarily be so. The ideal would be a zero cost design decision which enabled a collector to be installed at any future time at a cost close to that of installing it in a new construction' (15).

Another reason why designers need to have in mind the possibilities for change is to enable full advantage to be taken of technological innovation. For example, solar collectors are now being developed which claim efficiency values of 61% under cloudless sky, 58% in hazy weather, 45% when slightly overcast and still 20% when heavily overcast. This compares with 53%, 46%, 17% and 0% performance of the usual flat collector (16).

The cost of improving the thermal insulation of a building, once it has been built, can be very high. The penalty for establishing the level of insulation too low at the outset may thus also be very high. Assuming a choice between two kinds of wall construction with equal U-value and cost in which one is solid and the other has a cavity, the preference should obviously be the cavity wall since this allows insulation to be readily improved later. In situations where options cannot be left wholly open, paying for additional insulation at the outset has to be seen as an insurance premium to cover the effect of the 'irreversibility' of the decision.

In terms of fenestration, the most dramatic possible demand in the future might be to remove sealed windows and replace them with openable lights. Of course, this is possibly an extreme case, but could be necessary if the cost of air-conditioning became prohibitive. If options were to be kept open, therefore, it would be prudent to design sealable opening windows if the advantages of so doing outweighed any extra cost. One of the advantages could be the feasibility of switching to air conditioning only in more inclement seasons. It may incidentally be prudent anyway to cover periods of air-conditioning plant maintenance and replacement without totally interrupting activities. In developing countries it may be essential, to cope with electrical supply failure, to have alternative sources of fresh air ventilation.

Shading devices are perhaps the most readily modifiable elements and it would be easy to envisage the installation of extra non-structural shading on to a completed building.

ENVIRONMENTAL STANDARDS

Standards of environmental comfort have been improving steadily. In 1965, the Pilkington Research Unit stated that 'Environmental provision either stands still or improves (it goes back only during industrial "dark ages") so it is reasonable to assume that what is now exceptional will eventually become general' (17). The situation today looks very different, and the assumption that to be bigger or brighter or warmer is better needs critical examination. Standards of lighting of 1000 lux or more, which are frequently adopted today, are not necessary to meet task lighting needs. Waste heat from the lamps has to be extracted and engineers have to perform 'acrobatics of design in trying to make the air-conditioning appear to be designed with regard to fuel economy' (18). High overall intensities of illumination are not generally needed and artificial lighting is itself appallingly inefficient in its energy utilisation, fluorescent lamps are less than 20% efficient in their use of electrical energy. As Knight points out the saving in fuel and power that would accrue from more efficient lighting could warrant a total re-direction of scientific and development effort (19).

Thermal comfort standards too have a considerable impact upon energy use. Restricting internal winter conditions to 20°C in commercial buildings helps to reduce energy needs by about $\frac{1}{20}$th and similar savings are possible in summer. The standard internal design conditions in the UK are 21°C when the outside temperature is at 28°C. If the internal conditions were allowed to rise to 22°C the cooling energy would be reduced by about one-sixth.

If Government is prepared to establish mandatory standards for internal insulation in order to conserve energy, there is no logical reason why it should not also restrict illumination levels and establish other comfort criteria. The General Services Administration in the USA, the counterpart of the UK Property Services Agency, has produced guidelines for designers which, for example, establish maximum fuel and energy use of 173·5 kWh/m^2 floor area per year including fuel, light and power. Lighting and thermal comfort standards consistent with this rule have been suggested.

The possibility of reducing some environmental comfort standards to save fuel highlights how little is known about what comfort really is. Standards are at best a simple codification of scientific knowledge and rarely do justice to the variations of human response or to the interdependence of environmental effects. While we sense that people care about the direct control they have over their immediate environment the trends are still towards impersonally established comfort conditions. How much is known about the combined effects of heat, light, noise and humidity on personal comfort and mental performance? Can extremes of one factor be compensated for by the others? Environmental standards have advanced along their separate paths; calling them into question may enable a new and more comprehensive attempt to be made to define comfort in terms of human experience and expectation.

CONCLUSIONS

Having spent some time considering the future it might appear risky to offer further speculation, yet the futures for design vary so widely that merely to define the extremes may be helpful.

On the one hand, if the trends established by the USA are followed, and energy consciousness is backed by Government decree independent of market forces, we may find a move towards a lower technology design and maintenance, with a focussing of high technology solutions upon activities with severe environmental requirements.

On the other hand, if new sources of energy are found to supplement the known reserves then the price mechanism alone will allow 'high quality' environmental control to all who can afford it.

Perhaps the direst crisis gives rise to the most paradoxical outcome; if energy conservation really does become the sole criterion for human endeavour we shall end up in air-conditioned buildings, but underground, reducing thermal transmission to the minimum (20). This surely is a most chilling prospect.

In considering orientation and building form, building elements and environmental standards, some of the factors have been identified that influence the architect's design decisions related to air-conditioned buildings today. Further, in recognising the uncertainty of future fuel price and availability, an effort has been made to indicate how design decisions can be made less irreversible and can thus more readily accommodate modifications in the future.

REFERENCES

1 G. Brown and P. Stellon, The Material Account, *Built Environment*, August, 1974, pp 415–417.
2 Several analyses of the uncertainties facing organisations have been published. A distinction has been suggested between—
   (i) uncertainty arising from a lack of knowledge about future trends
   (ii) uncertainty arising from the unknown future intentions of others and
   (iii) uncertainty due to value judgements.
   By tracing through the consequences of uncertainty upon the outcome of a decision it is possible to assess the sensitivity of a decision to particular assumptions. A robust decision, in this context, is one which will retain flexibility to accommodate future contingencies.
   J. F. Friend and W. N. Jessop, *Local Government and Strategic Choice*, London, Tavistock, 1969.
3 L. March, *Some Elementary Models of Built Form*, Department of Architecture, University of Cambridge, Land Use and Built Form Studies Working Paper No 56.
4 ibid, p 61.
5 ibid, p 61. The thermal image of a surface is the mapping of the total heat flow through the surface and depends upon both the area and the thermal transmission characteristics.
6 R. W. Marks, *The Dymaxion World of Buckminster Fuller*, Reinhold, 1960, pp 120–133.
7 J. C. Knight, Engineering Economics of Energy Conservation. Paper given to Conference on Energy and Buildings – Towards a New Policy, 30 April 1974, p 1.

8 YRM International for Department of Health and Social Security, *Health Buildings in Hot Climates*, HMSO, 1977.

9 P. Madan, Design for Indoor Comfort, *Middle East Construction*, May, 1977.

10 A. C. Hardy and P. E. O'Sullivan, *Insulation and Fenestration*, Stocksfield, Oriel, 1967, p 27.

11 ibid, p. 29.

12 In double glazing units glass panes take on the temperature of the adjacent environment. The heat resistance is formed by the air entrapped in the space. Apparently one-third of the heat transfer between the panes is by convection, while two-thirds takes place through radiation. In applying a heat reflecting layer onto the glass in form of tin oxide or Indium oxide, radiation can be reduced drastically. The light transmission is hardly impaired by selection of a suitable thickness of the deposit. The convection in turn can be reduced by the introduction of a heavy noble gas like Kripton or similar.

13 A. Schneider, Energie und Raumklima, *Ansatzpunkte fuer eine neue Architectur*, Bauen & Wohnen, 8, 1977.

14 eg S. Carr, The City of the Mind, *Environment for Man: The next 50 years*, (ed) W. R. Ewald, Indiana University Press, 1967, pp 197–228; Society and Its Physical Environment, *The Annals of the American Academy of Political and Economic Science*, May, 1970; H. M. Proshansky, W. H. Ittelson, L. G. Rivlin, (Eds) *Environmental Psychology: Man and his Physical Setting*, New York, Holt, Rinehart and Winston, 1970.

15 D. J. Fisk, *Energy Conservation: Energy Cost and Option Value*, Building Research Establishment, DOE CP 57/76, 1976, p 4.

16 The solar collection in the panel consists of evacuated glass tubes housing an absorber in the form of tubes through which water is carried. A filtering layer on the top combined with a mirroring silver on the lower half of the tube prevents re-radiation onto the absorber, providing thus a system with a much better performance.
The Experimental House, Philips Publication Ref 17. 9861. 02. 4. 667. 11.

17 P. Manning, (Ed) *Office Design: a Study of Enviroment*, Pilkington Research Unit, Department of Building and Design, University of Liverpool, 1965, p 20.

18 Knight , *op cit*, p 6.

19 ibid, p 7.

20 Several studies have been made of the potential of underground space in an energy scarce society. While indicating that improved insulation for above-ground construction cannot begin to compete with under-ground construction in terms of energy conservation, one report claimed that 'there appear to be no adverse psychological effects of working in properly designed underground buildings'. While preposterous in terms

of the evidence produced to support it, the claim is unfortunately typical of the environmental determinist viewpoint.

T. P. Bligh and R. Hamburger, Conservation of Energy by Use of Underground Space, *Legal, Economic and Energy Considerations in the Use of Underground Space,* National Academy of Sciences, Washington DC, 1974, pp 103–118.

Even more bizarre is the assertion that 'a compact underground city (is) the end point of the spectrum of options between the present energy abundant life styles and the most energy constrained system'.

Technology of Efficient Energy Utilisation, *The Report of a NATO Science Committee Conference,* Les Arcs, France, October 1973.

ACKNOWLEDGEMENT

The author gratefully acknowledges the help of David Holland, Franz Levi and Michael Cassidy, his colleagues at YRM.

DISCUSSION

L. J. Wild (George Wimpey & Co Ltd)   Although not agreeing with the concept Mr Allford nevertheless suggests that thermally the optimum solution is for buildings going deep into the ground. I do not agree that this would be an optimum. It is very expensive to build downwards, there is a large energy use carrying out the excavation and fire protection for a structure below ground would be much more expensive with present regulations requiring a four hour rating against the two hour rating required above ground. Buildings may also have to be larger because a greater proportion of the area of the building will be required to provide means of escape from these large holes.

D. Allford (Yorke Rosenberg Mardall)   I agree with what Mr Wild says. Building underground was mentioned because there is published American work which refers to totally underground buildings in Texas. It was included partly to be provocative, as it is in our view a rather anti-human concept, although the technology is available to do almost anything. I have seen work on housing by an Australian architect where the ground had been scooped and mounded – rather cheaper than excavation – and low technology solar cells and windmills etc, used for energy, related of course to the Australian climate. The housing was on two storeys, blocked by the ground at the back and with little courtyards cut into it at the front.

5 Energy optimisation in the design of an air-conditioned building

T. SMITH
V. A. HAMMOND

INTRODUCTION

Any air-conditioning system is a composite of many component parts, all of which, in their manufacture require the expenditure of energy and many of which in their operation consume or release energy. Responsible manufacturers specify their components in a variety of ways which can be interpreted in energy terms, for example, the power characteristics and efficiency of pumps, fans, refrigeration machines, etc. From this data the designer can select components which individually optimise the use of energy. When, however, the component parts come together to form a complete air-conditioning system the total result is not the simple summation of individual parts. There is no simple method for optimising the total system in energy terms, the process is refined and extensive. It is impossible to rationalise system choice without relating the system to the building it is to serve, as aspects of the building such as orientation, massing, structural form, fenestration, etc influence many aspects of the air-conditioning system. Lastly it is equally impossible to rationalise without taking into consideration the artificial lighting installations within the building which also react with the system.

This chapter sets out to relate an experience of rationalisation and uses as its base a building which has been designed and constructed but not yet tested in its total operation. It should not be considered as a case study, which term is generally applied to completed and tested constructions, but we feel in presenting it that more is to be learnt from fact than from theoretical fancy. The building forming the basis of this chapter is the new Central Electricity Generating Board Headquarters Building in Harrogate, Yorkshire, UK. While the rationalisation processes described are applied to an Office Building they are equally applicable in their form to any building constructed for any purpose. It is not claimed that the building which has evolved is one which, in energy terms, is totally efficient. It could be claimed to be a fair compromise between the requirements of the accountant, the scientist and the user in that aspects of strict thermal efficiency and maximum energy conservation were compromised to achieve a building in financial terms acceptable, and in human terms agreeable. It is our belief that all such scientific design should be so compromised.

THE BRIEF

The CEGB issued a comprehensive brief to their professional team, from which the following extracts are quoted:

'The site is a magnificent choice, but if it is to bring credit to the Board and also provide enjoyment for the staff, it will require sensitive development'.

'The new building is to be built in a period of acceleration in cultural change which should affect what is to be built; it will be asked to do different things; the new needs, eg open planned offices demand new forms which in turn bring unfamiliarity'.

'Unfamiliarity is always disturbing and there is already a general awareness that as buildings become more mechanically refined they become more inhuman, and that while occasionally they excite or surprise, they give no reassurance, sense of place, and become depersonalised through size and repetition'.

Energy conservation was implicit in the brief. However, it was emphasised that the building should not be extreme in any way and that the high quality of the built environment, internally and externally, was a major parameter in establishing the designs.

The CEGB supplemented their brief so that

• the building should incorporate low energy characteristics and provisions, so far as they were consistent with the general stipulations on the quality of the environment, and

• that all provisions specifically incorporated to reduce energy consumption should be viable financially in terms of capital expenditure and savings in energy costs.

Interpretation of brief

In 1969 the Electricity Council published a booklet entitled *Integrated Design – a case history*. It described the co-ordinated design processes involved in the Testing and Research Station at Wallsend, owned and operated by the NEEB. They included co-ordinated investigations to optimise detailed financial aspects of the project, and special studies of:

• size and shape of building
• 'U' factors of structural elements
• areas and types of fenestration
• natural and artificial lighting
• heating, cooling and ventilation
• acoustics.

The technique involved has been developed and used in the design of subsequent buildings and is now generally known as IED – Integrated Environmental Design.

It has been defined (by the Electricity Council) as a concept and process of design applied to buildings and their services providing a practical approach to optimising three factors.

(1) The quality and consistency of internal environment in relation to the functions of the building and the needs of the people using it.
(2) Capital costs.
(3) Running costs.

High rates of heat recovery and low building thermal balance temperatures have received much publicity over recent years. However they are not a reliable index of energy conservation. Too often they have been achieved by virtue of unnecessarily high initial energy input for lighting and other purposes.

Whilst IED procedures emphasised cost implications, they involved consideration of energy consumption. The procedures could be amended so that they emphasise energy consumption, as a direct indication of potential energy conservation.

Energy input to buildings relates primarily to three factors:
(1) The permeability of the building envelope to heat and light.
(2) The energy input necessary to operate services to counteract the effects of (1).
(3) The energy input necessary for adequate artificial lighting.

It follows that in establishing the nature and proportions of the building elements, their merits have to be assessed against the energy requirements arising because of them. Secondly, the services designs have to be assessed against the energy requirements necessary for them to function.

A factor in the total energy requirements for services is the extent that residual heat from the expenditure of electrical energy for lighting and other purposes can be recovered, upgraded if necessary, and re-used.

The transfer of energy from one medium to another inevitably results in energy loss due to the inherent inefficiency of transfer equipment. The concept of elevating lighting levels above that required for optical function in order to create excess energy which is subsequently recovered inefficiently and re-used was therefore considered unacceptable.

It was decided that methods of analysis developed for IED should be adopted and adapted as necessary to ensure that the client's brief was interpreted conscientiously, particularly with regard to energy conservation.

In addition to adopting modified IED techniques, considerable use was made of the Central Electricity Council's BEEP programmes and adaptations thereof. The investigations followed the general pattern and scope as described for IED, *but re-orientated with the definite objective of minimising primary energy input and maximising re-cycling of surplus residual heat.*

The emphasis on low energy consumption rather than on a 'self-heating' building without a significant heating plant was a predominant theme of the co-ordinated building designs, leading to the maximum in energy conservation consistent with the requirements of the client's brief.

THE BUILDING

Original concept

The first architectural sketch plans indicated a complex shape on three floors with part basement. They indicated a building with a fairly light structural frame, insulated decking roof, and floor to ceiling fenestration. The proposal was for a building that could be constructed at reasonable cost and within a reasonable period.

The intentions were that the building would not be intrusive in the landscape and would provide maximum facilities to enjoy the outward view through the windows.

The glazing was to be of a reflective type so that the building reflected its surroundings and merged into the landscape.

Preliminary design assessment

The first investigations were to establish whether a building with an acceptable annual energy demand could be constructed within the broad terms of the preliminary architectural concept and inevitable financial limits imposed in capital terms.

A series of maximum values for thermal performance were established which were:

(1) The ratio of envelope area to building volume combined with the overall average thermal transmittance (U) value should be such that the total mass air flow rates should be not greater than
  (a) 10 air changes per hour in perimeter zones, and
  (b) 8 air changes per hour in interior zones
  with a maximum temperature differential between supply air and room of 9°C. The maximum differential occurs on the cooling cycle.

(2) Since the initial plans included maximum glass areas, for visual and psychological amenity, it must be of sealed double glazed type in insulated frames to ensure a satisfactory internal environment.

(3) The total area of unshaded, clear, sealed double-pane glazing should not exceed 30% of the floor-to-ceiling wall area on any one elevation.

(4) If suitable shading devices were incorporated, or if double-pane solar control glass were used, the glazed area could be increased proportionately to the reduction in solar transmission.

(5) The aspect ratio of the external walls should be 1:1 if feasible. The ratio should not exceed 1.5:1.

(6) The external walls (excluding glazing) should have an average U value not greater than 1.0 W/m² °C.

(7) The roof (excluding ceiling void and ceiling) should have a U value not greater than 1.0 W/m² °C. The external surface should be a light colour (eg stone chips) for maximum reflectance.

The building as defined in the sketch plans did not conform to these parameters. Further studies were initiated so that suitable modification could be effected.

Building planning
The sketch plans were modified and developed progressively with checks against the design parameters between each series of changes. The ratio of envelope area to building volume was reduced progressively until a building plan acceptable to the whole design team was achieved.

The final design comprises a building 86m long and 75m wide (an aspect ratio of 1.115:1). It is three storeys high. There is an open central court in the middle, which forms a landscaped water garden.

The main axis of the building was always approximately east-west. This was not the optimum when considered in conjunction with earlier plans, but became more acceptable as the building progressed to an approximately square plan. The eventual building orientation locates the 86m elevations of the building 15° west of north and 15° east of south.

The building is located on high ground, very exposed and with no protection or shading.

The final shape is not the ultimate from the aspect of energy consumption. It would have been better as a higher building with no internal court. However, such a building would not have fulfilled criteria relating to sensitive development, and would not have been so satisfactory for its intended functions and for administration. Furthermore, the height of the building was limited by restrictions in the planning consent.

The final plans provide a total floor area of 19,600 m², 16,400 m² of which is on the three main floors and 3200 m² in the partial basement. Due to the

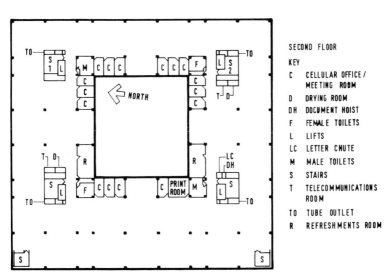

5.1 Typical office floor plan

natural slope of the site and planned excavations it is now a lower ground floor with direct access from outside at ground level.

The lower ground floor accommodates plant rooms for heating and cooling plant, emergency diesel generator, ventilation plant, telephone rooms, water storage, electrical intake and switchrooms, car park, stores and a small leisure pool.

The ground floor is of compartmented design. It houses the main reception and display areas, library, lecture theatre, medical suite, secretarial and reprographic departments, and restaurant and coffee areas, and ventilation plants that serve only the ground floor.

The first and second floors are mainly open plant with small areas of compartmented offices around the central courtyard.

Fig 5.1 illustrates the second floor layout.

Fenestration
An all-glass building facade as envisaged in the early architectural studies was untenable in terms of energy consumption needed to offset the heat gains and losses involved. In addition, a mass air flow rate in the order of 18 air changes per hour in perimeter zones would have been needed. This exceeded the maximum rate of 10 changes allowed in the original provision, and entailed further energy penalties.

In order to establish the preliminary performance parameters equivalent to 30% of clear, double-pane glass, an examination was made of existing buildings and research data. At the time, it was appreciated that larger areas would be essential to permit maximum enjoyment of the views of the surrounding countryside.

Further studies were initiated to establish maximum glazing area and types with conservative energy requirements and acceptable costs. The studies included solar control glasses from UK and abroad and the types of glass covered the following range:

Glass type	tinted and clear, double pane and sandwich, reflective and absorbent, shaded and unshaded
Daylight transmittance	87% to 18%
Solar transmittance	84% to 16%
Shade factor	0.97 to 0.3
U value	5.6 to 2.1 W/m² °C

The specification eventually selected was:

Glass type	tinted, reflective, bronze, double pane
Daylight transmittance	33%
Solar transmittance	25%
Shade factor	0.28
U value	2.1 W/m² °C

Tables 5.1 and 5.2 are actual design studies relating to the fenestration that was included.

Table 5.3 is a synopsis of the results of the principal studies that were

Table 5.1   Fenestration – instantaneous maximum demand

Heating and Cooling loads

Appendix No. 13
Based on Pilkington *Sun-cool, bronze 35/25*
Double glazed windows 50% of walls
Lighting *750 lux*

Building System load elements	Maximum heat loads		Maximum running load (Peak energy input)	
	Winter kW	Summer kW	Electricity kW	Boiler fuel Oil l/h
1st and 2nd Floors (offices)				
Lighting	0	− 462		
Windows	+ 64	− 90		
Other sources	+562	− 649		
Total	+626	−1 201	353	
Heat reclaim	−585	0		
Net total	+ 41	−1 201		4
Ground floor—All day occupation (offices, services, stores etc)				
Lighting	+ 0	− 63		
Windows	+ 15	− 24		
Other sources	+103	− 98		
Total	+118	− 185	54	
Heat reclaim	−114	0		
Net total	+ 4	− 185		0.4
Ground floor – Transient occupation (restaurant, lecture rm, meeting rms etc)				
Lighting	0	− 29		
Windows	+ 12	− 17		
Other sources	+317	− 219		
Total	+329	− 265	78	
Heat reclaim	− 19	0		
Net total	+310	− 265		29
Mechanical services plant				
Air handling units – offices			50	
Air handling units – GF All day			10	
Air handling units – GF Transient			14	
Fans – car park			4	
Boilers			8	
Pumps			40	
Cooling towers			20	
GRAND TOTAL	+355	−1 651	631	33.4

Table 5.2   Fenestration – average annual energy, consumption

Approximate energy consumption	Appendix No. 14
	Based on Pilkington *Sun-cool bronze 35/25*
	Double glazed windows 50% of walls
	Lighting *750 lux*

Building zone Systems	Maximum instantaneous Electrical load		Equivalent full-load hours h/annum	Annual energy consumption MWh/annum
	Winter kW	Summer kW		
1st & 2nd floors (offices)				
Lighting – perimeter	112	0	550	62
Lighting – interior	350	350	2 500	875
Refrigeration – all year	147	147	2 150	316
Refrigeration – summer	0	206	1 040	214
Air handling units	50	50	2 500	125
Total	659	753		1 592
Ground floor (office, stores, restaurant, lecture rm, meeting rms, etc)				
Lighting – all day	56	56	2 500	140
Lighting – transient	45	45	1 250	56
Refrigeration – all day – Summer	0	54	1 040	56
Refrigeration – all day – Winter	30	0	800	24
Refrigeration – transient Summer	0	78	520	41
Refrigeration – transient Winter	12	0	250	3
Air handling units – all day	10	10	2 500	25
Air handling units – transient	14	14	1 250	18
Total	167	257		363
Basement car park				
Lighting	14	14	2 000	28
Fans	4	4	3 000	12
Total	18	18		40
Central plant				
Boilers	8	0	1 060	8
Pumps	40	40	2 500	100
Cooling towers	0	20	1 040	20
Total	48	60		128
Grand total	829	1 088		2 123
Boiler plant fuel oil consumption	33.4 l/h		1 060	35 400 litres/ annum (7 800 gall/ annum)

Table 5.3 Fenestration – a synopsis of the principal design studies

Study appendix no.		Maximum heat loads		Maximum instantaneous electric load		Annual energy consumption		Annual running cost	
		Winter kW	Summer kW	Winter kW	Summer kW	Electricity MWh/annum	Oil l/annum	Electricity £	Oil £
1+2 Tinted Glass Windows (equivalent to 30% clear single glass)	500 lux	609	1 547	718 150	891 150	1 758 176	61 500		2 160
3+4 Clear Glass Windows (equivalent to 30% clear single glass)	750 lux	421	1 702	868 929 150	1 041 1 091 150	1 934 2 164 176	43 500	27 040	1 520
5+6 Clear double glass windows with external louvre blinds	750 lux	475	1 608	1 079 923 150	1 241 1 076 150	2 340 2 141 176	48 760	32 380	1 720
7+8 Varitran Silver 1.120" double glazed windows 70% of walls	750 lux	427	1 763	1 073 923 150	1 226 1 207 150	2 317 2 383 176	43 460	32 110	1 540
9+10 Sun-Cool Azure 37/29 double glazed windows 65% of walls	550 lux	559	1 546	1 073 722 150	1 357 939 150	2 559 1 736 176	55 500	34 370	1 952
11+12 Antisun float grey 37/48 double glazed windows 65% of walls	550 lux	626	1 646	872 722 150	1 089 969 150	1 932 1 787 176	63 000	27 235	2 224
13+14 Sun-Cool Bronze 35/25 double glazed windows 50% of walls	750 lux	355	1 651	872 892 150	1 119 1 088 150	1 963 2 123 176	35 400	27 573	1 250
				1 042	1 238	2 299		32 290	

WINDOW TYPE	CLEAR	CLEAR	CLEAR	VARITRAN 1·20	SUN-COOL 37/29	ANTISUN 37/48	SUN-COOL 35/25
N° OF PANES	1	1	2	2	2	2	2
SHADING	TINTED	NIL	EXT LOUVRE	REFLECTIVE	REFLECTIVE	ABSORBANT	REFLECTIVE
PROPORTION OF FACADE	30 %	30 %	70 %	70 %	65 %	65 %	50 %
LIGHTING	500 LUX	750 LUX	750 LUX	750 LUX	550 LUX	550 LUX	750 LUX

5.2 Effects of windows and lighting on peak heating and cooling loads

WINDOW TAPE	CLEAR	CLEAR	CLEAR	VARITRAN 120	SUN-COOL37/29	ANTISUN 37/48	SUN COOL 35/25
N° OF PANES	1	1	2	2	2	2	2
SHADING	TINTED	NIL	EXT. LOUVRE	REFLECTIVE	REFLECTIVE	ABSORBANT	REFLECTIVE
PROPORTION OF FACADE	30 %	30 %	70 %	70 %	65 %	65 %	50 %
LIGHTING	500 LUX	750 LUX	750 LUX	750 LUX	550 LUX	550 LUX	750 LUX

5.3 Effects of windows and lighting on annual energy consumption

Table 5.4 Roof lights – instantaneous maximum demand and annual average energy consumption

Heating and cooling loads

Schedule D	Capital costs	
Roof lights 4 No. 1.8 m × 1.8 m	Lighting	+£5 400
per 10.8 m × 10.8 m module,	Refrigeration	−£2 300
with sloping glass. 500 lux	Boilers	+£1 200
	Total	+£4 300

Building zone	Maximum heat loads		Maximum running load (Peak energy input)	
System load elements	Winter kW	Summer kW	Electricity kW	Boiler fuel Oil l/h
Lighting deficit	+33.8	−27.9	−8.2	+3.2
Roof lights	+10.1	+ 1.5	+ .4	+1.0
Net effect	+43.9	−26.4	−7.8	+4.2
			(Saving)	(Extra)

Approximate energy consumption

Building zone	Maximum instantaneous electric load		Equivalent full-load hours h/annum	Annual energy consumption MWh/annum
	Winter kW	Summer kW		
	Saving	Saving		Saving
Lighting winter	27.9		600	16.74
Lighting summer		27.9	800	22.32
Refrigeration – lights	8.2	8.2	1 400	11.48
Refrigeration – roof lights	0	− .4	1 040	− 0.42
Net total electrical energy	36.1	35.7		50.12

	Maximum instantaneous boiler loads		Equivalent full load hours h/annum	Annual fuel oil consumption litres/annum
	Winter l/h	Summer l/h		
				(Extra)
Lighting deficit	3.2	0	600	1 920
Roof lights	1.0	0	1 060	1 060
	4.2	0		2 980
				(660 gallons/ annum)

made. Many supplementary studies were also prepared.

It is interesting to note that the artificial lighting intensities used in the studies vary due to concurrent studies on this aspect. It will be seen that the glass area is varied in each case to a value within the original design parameters whilst seeking large areas for visual amenity.

Figs 5.2 and 5.3 are histograms based upon the data in Tables 5.1, 5.2 and 5.3 to assist in the assessment.

Fig 5.2 illustrates that, although the window areas might be considered generous, the losses and gains are not unreasonable proportions of the total heating and cooling loads. The effects of fenestration vary with plan area and plan aspect ratio. It was for this reason that the studies were based upon total thermal loads for the building rather than the thermal and solar characteristics of a unit area of combined wall and window.

Walls

The building is of reinforced concrete construction with spandrel walls. The composite wall construction comprises external anodised aluminium panels, 25 mm of mineral wool insulation in polythene bags, a 50 mm air space, 110 mm of medium density blockwork, 13 mm of cement render and an internal finish. The overall U value of the composite wall structure is 0.80 W/m² °C. This compares with the preliminary target figure of 1.00 W/m² °C.

Roof

The early architectural sketch plans incorporated an insulated light decking roof. As the designs developed it was decided to locate a large number of fairly heavy air handling units on the roof (see later section) which led to a decision to abandon the light roof deck for a heavier construction.

After completion of structural studies, a heavy, insulated concrete roof was adopted. It comprised 0.325m of concrete, on top of which are 0.075m of insulation, covered with felt, asphalted and covered with light coloured reflective chippings.

The overall U value of the composite roof structure, excluding the ceiling void and suspended ceiling, is 0.83 W/m² °C. This compares with the preliminary target figure of 1.00 W/m² °C.

Roof lights

Just as the central courtyard provided visual relaxation for staff and increased natural daylighting, it was thought roof lights might achieve similar results. To investigate the viability of roof lights, four studies were made.

The first indicated that if roof lights were to be provided they would have to be of northlight type to minimise solar gains, of fairly heavy construction to provide inertia and double glazed to minimise losses and condensation.

The second study indicated that fairly small roof lights spaced at frequent intervals would provide a better overall energy performance than large roof lights at greater spacings.

Table 5.5   Roof lights – a synopsis of the principal design studies

Study of roof lights	Element	Maximum heat loads		Maximum instantaneous electrical load		Annual energy consumption		Annual running cost	
		Winter kW	Summer kW	Winter kW	Summer kW	Electricity MWh/annum	Oil l/annum	Electricity £	Oil £
1. 6 No. 1.8 m × 1.8 m per module	Lights	33.00	−27.28	−35.30	−35.30	−31.35	1 200		
Vertical glass	Roof lights	12.93	2.15	0	0.63	− 8.54	1 300		
North facing Lighting 500 lux	Total	45.93	−25.13	−35.30	−34.67	−39.89	2 500	−560	88
2. 6 No. 1.8 m × 1.0 m per module	Lights	16.50	−13.64	−13.64	−13.64	−15.68	600		
Vertical glass	Roof lights	12.93	2.15	− 4.01	− 3.38	− 3.95	1 300		
North facing Lighting 750 lux	Total	29.43	−11.49	−17.65	−17.02	−19.63	1 900	−275	67
3. 1 No. 3.6 m × 3.6 m per module	Lights	43.13	−35.65	−35.65	−35.65	−30.12	775		
Vertical glass	Roof lights	13.51	2.25	−10.5	− 9.84	− 8.18	1 357		
North facing Lighting 750 lux	Total	56.64	−33.40	−46.15	−45.49	−38.30	2 132	−540	75
4. 4 No. 1.8 m × 1.8 m per module	Lights	33.80	−27.90	−27.90	−27.90	−39.06	1 920		
Vertical glass	Roof lights	10.10	1.50	− 8.20	− 7.80	−11.06	1 060		
North facing Lighting 500 lux	Total	43.90	−26.40	−36.10	−35.70	−50.12	2 980	−703	106

NB Negative values denote savings

Two designs were prepared with plan dimensions of 1.8m x 1.8m and with approximately 2.0m² of double pane glass in each. In one design the glass was vertical and in the other it was inclined 60° to horizontal. It was possible to locate about 60 such units on the roof.

Detailed studies were undertaken to assess the probable performance of these alternatives. The inclined glass unit proved more efficient, and the results indicated.

(1) A daylight factor of 2% could be achieved over most of the floor. Daylight calculations based upon an overcast sky (CIE 5000 lux) indicated that natural lighting would provide a maximum illuminance of 700 lux and an illuminance exceeding 200 lux for 1400 hours per annum.

(2) Natural lighting could supplement artificial lighting for 1400 hours per annum. Accordingly some of the light fittings could be switched off for this period, a saving in energy input.

(3) The maximum instantaneous cooling load was not affected significantly, but the annual load was reduced by virtue of the reduction in heat dissipation from luminaires. This is a further saving in energy input.

(4) In winter, the energy saved by switching off lights reduces the quantity of heat reclaimed and re-cycled. Consequently, during the colder part of the winter, the amount of supplementary heating needed would increase. This is an increase in energy input.

(5) As an energy conservation provision, the roof lights would effect a reduction in overall energy consumption.

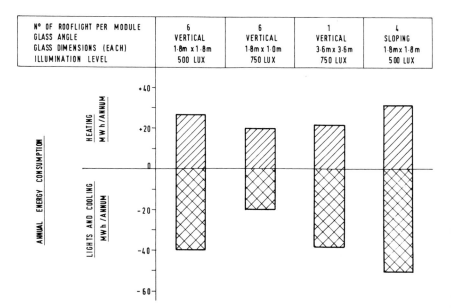

5.4 Roof lights – effects on energy consumption

The financial implications were also favourable and unit type roof lights were incorporated into the designs.

Table 5.4 illustrates one actual study showing the results of the inclusion of roof lights, the consequent reduction in energy for lighting and increase in supplementary heating.

Table 5.5 is a summary of the four principal studies. Again they indicate the influence of concurrent studies. The artificial lighting intensities incorporated are 500 and 750 lux. Supplementary heating is based upon oil, which seemed likely at the time, although electricity and gas were also under consideration.

Fig 5.4 is a histogram based upon the data in tables 5.4 and 5.5 to assist the assessment.

The roof lights analysed in Table 5.7 were eventually adopted – in conjunction with 750 lux illumination and gas fired boilers.

AIR CONDITIONING

Initial consideration

Three functions in particular are required in an air conditioning system:
- to provide adequate ventilation
- to provide adequate cooling
- to provide adequate heating.

To these three functions a fourth has been added of paramount importance.
- to provide adequate ventilation, cooling and heating with the lowest viable energy consumption.

A major element of total energy consumption is that necessary for operating the air-conditioning system. As required by the overall design concept for the building, attention was directed towards this element of energy demand.

Data was obtained on the types of air handling and distribution systems that had been used in publicised 'energy conscious' buildings constructed in recent years. There was no consistency in the systems used. They included combinations of:
- Single duct low velocity with centralised air handling plant.
- Single duct low velocity with de-centralised air handling plant.
- Induction systems for both perimeter and inner zones.
- High velocity variable volume for inner zones and induction units around perimeter.
- High velocity variable volume for inner zones and variable volume with re-heat for perimeter zones.
- High velocity dual duct systems.

The data confirmed that many buildings incorporated large central air handling plants with extensive ranges of high velocity ductwork.

In such designs the ducted air is not only used to effect satisfactory air movement, temperature control etc in the conditioned space; it is also used as a medium to convey thermal energy between the conditioned space and the central plant. Because of its low specific mass and low specific heat capacity air is not a good medium for conveying thermal energy. Since the functions that affect the energy needed to transport air include velocity of mass flow, it therefore follows that more energy is needed at higher velocities.

Because of its higher specific mass and specific heat capacity compared with air, water needs less applied energy to transmit the same thermal quantity.

Table 5.6 is a summary of data abstracted from the Guide based upon 70 kW transported in air at low and high velocity and 10°C differential and water at normal velocity and 10°C differential (for hot water) and 5°C (for chilled water). Although the data is restricted to a length of one metre of straight conduit, and the differences would not be so dramatic for complete systems, it is indicative of the great variation in energy for motive power that could arise from system selection and design.

Another factor in energy required for operating air conditioning systems is the comparative power characteristics and operating efficiencies of pumps and fans of high and low capacity. Since the building was to be one with low energy consumption, it was decided that consideration based on these principles might provide the most viable solution.

Table 5.6 **Data abstracted from CIBS Guide relating to the transport of 70 kW of thermal energy in air and in water**

	Air	Water
Specific mass (density) – kg/m^3	1.205	1.000
Specific heat capacity – kJ/kg°C	1.012	4.187

	Air		Water	
Energy transferred – kW	70	70	70	70
Temperature difference – °C	10	10	5	10
Mass flow rate approx – kg/s	7	7	3.4	1.7
Volume flow rate approx – l/s	5 800	5 800	3.4	1.7
Velocity approx – m/s	5	20	1.0	1.0
Conduit diameter – mm	1 200	600	65	50
Resistance to flow (1) – N/m^3	0.2	6.0	142	153
Efficiency	65%	75%	75%	75%
Applied (motive) energy (2) – W/m	1.8	46.4	0.64	0.35

(1) N/m^3 = N/m^{2}m
(2) From CIBS Guide Section B.

$$\text{Watts} = \frac{\text{m}^3/\text{s} \times \text{N/m}^2}{\text{Efficiency}}$$

System selection

As stated, the building incorporated three floors and a part lower ground floor. It had been agreed that plant could be located in the basement and on the roof. (This was one of the decisions which contributed to a change in the roof from insulated decking to a heavier construction.)

Ceiling voids, of reasonable dimensions to house services, had been incorporated on all floors.

The building included four service cores which incorporated lifts, staircases, toilet areas etc. Provision was made for a vertical service shaft in each core passing through all levels of the building.

It was intended that services equipment needing floor space should be eliminated as far as possible.

Consideration was given at first to the first and second open plan floors. From the principles relating to motive power and from the locations and dimensions of shafts, voids and cavities provisionally allocated, it was decided that:

(1) With one central air handling plant, a high velocity system was essential to maintain reasonable services space provisions. Total motive power would be at a maximum.

(2) With two, or four, air handling plants associated with the building cores, high velocity ductwork would still be needed, but the total motive power would be reduced.

(3) A further increase in numbers of air handling units suitably dispersed over the roof, each located as close as possible to the area it served, permitted the use of short, simple, low velocity ductwork systems that could be housed within the proposed ceiling voids. Further reduction in total motive power would be achieved. Greater flexibility of use by switching individual units was possible with inherent savings in energy consumption.

It was decided to adopt the principle of multiple air handling units with low velocity ductwork.

Even though the fenestration was to be of double-pane type, it was appreciated that provision must be made to minimise down-draughts and cold radiation in order to create an acceptable environment around the perimeter. Investigations indicated that this could be achieved with properly designed and located linear diffusers around the perimeter and mounted in the ceiling.

The compartmented areas of the ground floor were next considered. It was decided that the most viable solution would be to provide multiple air handling units in plant rooms at the same level. The numbers and functions of the units were related to the various departments and operational schedules. This also permitted the use of low velocity ductwork systems and the solution is compatible with that for the first and second floors.

Heat reclamation

It was decided that the methods of heat reclamation should follow well established principles:

- return air from conditioned spaces would pass through ventilated luminaires and thence through ceiling voids and return air ducts to the air handling units
- the refrigeration plant would be used for reclaiming and re-cycling heat, through a condenser circuit
- supplementary heating plant would be provided to offset any deficiency between heat reclaimed and heat demand.

In the pursuit of optimum energy conservation, the viability was examined of 'free cooling' from fresh air combined with heat recovery from exhaust air. Two studies were made, one utilising thermal wheels between fresh air supply and exhaust and the other incorporating chilled water heat recovery batteries in air exhausts.

Heat recovery and transfer between exhaust and fresh air on each individual air handling unit, by means of a thermal wheel, was not viable since:

- the cost was excessive
- space requirements were excessive
- apart from start-up, a large proportion of the air handling units would operate on a permanent cooling cycle
- each air handling unit would incorporate provision for 'free cooling' when outside temperatures are suitable.

It was apparent that although considerable quantities of heat could be reclaimed it could not be re-cycled effectively. If reclaim and re-cycling were to be viable, it had to be flexible and permit heat reclaimed from air handling units on cooling cycles to be transferred to air handling units on heating cycles. The most effective way to achieve this is through the central refrigeration plant.

To transfer heat from the exhaust to the fresh air intake of each air handling unit, and then remove it through the cooling unit is complicated and not particularly efficient. The controls are complicated and a lot of the advantages of 'free cooling' are lost. Accordingly this concept appeared abortive and possible supplementary studies incorporating run-around systems or plate-type exchangers were abandoned.

A second study was then initiated. It was based upon:

- the retention of a 'free cooling' facility on each unit to eliminate unnecessary refrigeration load
- a cooling coil in each air exhaust connected to the chilled water distribution system
- an automatic control cycle which at all times limited heat reclaim from exhaust air to the maximum that could be effectively re-cycled.

The analysis indicated that appreciable energy savings would result from these methods and they were incorporated into the designs.

The one exception to this general statement was the air conditioning plant for the leisure pool complex in the lower ground floor. In this case the most viable solution was the provision of a thermal wheel between supply and exhaust ducts. The thermal wheel reclaims a proportion of both sensible and latent heat from the exhaust duct.

Heat storage

The capital costs of providing special plant for storage of reclaimed heat are often excessive when measured against energy saved. Since the amenities of the building incorporated a small leisure pool (installed to satisfy the requirements for a static water supply) a study was made to establish the viability of using it as a heat store.

Since the pool was to be heated by a water-to-water calorifier connected to the heating distribution network, it was possible to elevate the pool water above normal temperature when a surplus of reclaimed heat was available. Heat could be removed from the pool and transferred into the overall reclamation and re-cycling system by a further water-to-water calorifier connected to the chilled water distribution network.

The study indicated that the system provided appreciable energy storage for re-use and that the reclamation cycle was of particular value under early morning start-up conditions.

It was decided to incorporate the heat storage facilities. Some of the data derived from the study is set out below.

Table 5.7

Reclaim capacity	
Pool content	= 135 000 litres
Maximum storage temperature	= 30°C
Reclaim low limit temperature	= 25°C
Chiller heat of compression factor	= 1.21
Reclaim capacity:	
135 000 kg × 4.187 kJ/kg × 5°C × 1.21	= *3.4 GJ*
Rate of reclaim	
Pool circulation rate	= 45 000 litres/hour
Calorifier temperature IN/OUT	= 30°–25°C
Chiller heat of compression factor	= 1.21
Reclaim rate:	
45 000 kg/h × 4.187 kJ/kg × 5°C × 1.21	= *1.14 GJ/h*
Heat reclaim period = 3 hours	
Rate of heat-up	
Pool circulation rate	= 45 000 litres/hour
Calorifier temperature IN/OUT	= 25°–28°C
Heat-up rate:	
45 000 kg/h × 4.187 kJ/kg × 3°C	= *0.57 GJ/h*
Recovery period = 6 hours	

System specification

Arising out of these studies it was possible to stipulate a design specification for the whole of the air conditioning, as follows:

(1)   The first and second floors would be air conditioned by multiple roof-top air handling units.

(2)   The ground floor would be air conditioned by multiple air handling units located on that floor.

(3)   Ductwork should be as short and simple as possible, with low velocity air delivered through ceiling mounted grilles.

(4)   Return air should be through ventilated luminaires, the ceiling voids being used as return air ducts where practicable, and should be re-cycled and re-circulated.

(5)   The mass air flow should be limited to that necessary for adequate heating, cooling and ventilation.

(6)   The minimum fresh air flow should not exceed that recommended in relevant sections of CIBS Guide.

(7)   Fresh air should be used for 'free cooling', coupled with heat recovery from exhaust air, when this contributes to minimum primary energy input.

(8)   Cooling should be provided by a central refrigeration plant with a chilled water pipework distribution to the outlying air handling units.

(9)   The refrigeration plant should be used for heat recovery and re-cycling. A closed condenser water circuit should form the heating pipework distribution to the outlying air-handling units.

(10) The recreational pool should be used to store, for subsequent re-use, surplus heat from the reclaim system.

(11) A heating plant should be incorporated to make up any deficit when the total heat reclaim is less than the heating load.

(12) Cooling towers should reject surplus when the cooling load provides unwanted heat in the condensers.

Design proceeded along the lines of the specification.

Fig 5.5 is a simplified diagram illustrating the significant aspects of the total design concept.

Air handling units

The air handling units incorporate normal equipment for fresh air, recirculation, filtration, humidification, heating and cooling. Each unit has either one or two zones and is of draw-through type. Each two-zone unit incorporates a common cooler battery and two heater batteries, one for each zone.

In two-zone units the relevant zones comprise identical areas on the first and second floors. Thus the periodicity and characteristics of the heating and cooling loads from the two zones are similar and the loads vary from each other only by virtue of the roof structure: this permits close control of heating and cooling battery outputs and is important in reducing mixing losses.

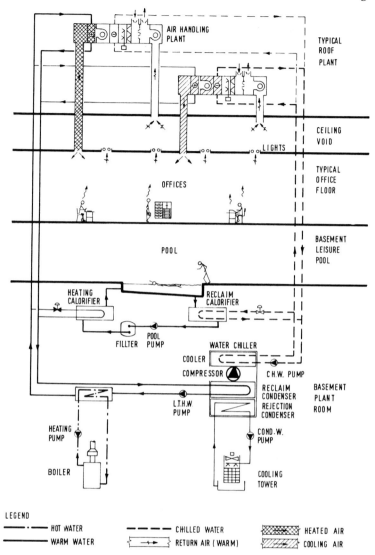

5.5 Diagram of heat reclaim system

Financial considerations dictated the desirability of using exposed water-proof air handling units on the roof. The number of units and their diverse locations over the roof area inhibited the provision of separate plant rooms or enclosures.

Proposals were obtained from a number of manufacturers of packaged air handling units. The most suitable proposal was then developed into a complete prefabricated, weatherproof air conditioning plant. The casing is of insulated, frame and panel construction, the panels being either

removable or in the form of hinged airtight doors for access and maintenance. Each unit is complete with ancillary equipment. Each unit incorporates a services compartment in which is housed all necessary electrical and
control equipment, prewired to the individual plant components.

Refrigeration plant
Studies were made of various types and combinations of refrigeration plant.
Whilst the primary purpose of the plant was to provide cooling, it was
important to optimise its performance as a heat pump on the reclaim cycle.
The selection of water temperatures on and off evaporator and condenser
also affect the design of the cooling and heating distribution networks and
the terminal heating and cooling equipment.

The equipment studied in combination included:
- centrifugal compressors with inlet guide vane capacity control
- helical screw compressors with slide valve capacity control
- single bundle shell and tube condensers with closed circuit cooling towers
- double bundle shell and tube condensers with open circuit cooling towers.

The combination selected comprises two refrigeration machines and two
open circuit cooling towers. The refrigeration machines are of centrifugal
type with double bundle condensers. They are connected together with
- evaporators in series
- heat reclaim condensers in series
- heat reject condensers in parallel
- cooling towers in parallel.

To control water quality and to limit the fouling factor of the heat
rejection condenser and its water circuit to the cooling towers, an automatic
filter treatment plant is incorporated. Its function is to remove solids, control
hardness and alkalinity, inhibit corrosion and to limit fungus, algae, etc.

Table 5.8   Refrigeration machines – performance analysis

Compressor/motor type and capacity control	Hermetic centrifugal with inlet guide vanes	Semi-hermetic helical screw with hydraulic by-pass slide valve control	Hermetic centrifugal with inlet guide vanes
No. of machines	2	2	2
Total cooling cap	498 Tons R	520 Tons R	491 Tons R
Cooler flow rate	65.86 l/s	65.86 l/s	65.86 l/s
,, press drop	54 kN/m^2	48 kN/m^2	54 kN/m^2
Reclaim flow rate	48.00 l/s	47.62 l/s	48.20 l/s
,, press drop	27 kN/m^2	48 kN/m^2	21 kN/m^2
Reject flow rate	48.00 l/s	2 × 23.81 l/s	47.90 l/s
,, press drop	11 kN/m^2	27 kN/m^2	21 kN/m^2
Power input (max)	490 kW	457 kW	456 kW
COP	3.56	3.81	3.82
	Selected plant		

The preceding table summarises some of the data relating to the types and combinations of refrigeration plant that were considered.

Although the selected machines had the highest maximum power input, they also had the lowest power consumption for ancillary pumps etc and the lowest overall annual average energy consumption.

Heating plant

The concentration upon minimum primary energy input that is a feature of the design limited the amount of residual heat available for reclamation and re-cycling. A supplementary heating plant was needed.

Three possibilities were examined.

(1) A low temperature hot water system operating at about 80°C flow and 70°C return, with gas-fired boilers.
(2) A similar system with oil-fired boilers.
(3) A similar system connected to electro-thermal heating and storage plant, on load at night, pressurised and with a maximum storage temperature of about 160°C.

Global fuel and energy problems that are still with us were very much to the fore when the studies were made, and complicated the selection of heating plant and fuel. Visions of a land flowing with North Sea oil were rampant. We were assured that supplies of Natural Gas would be plentiful when the Frigg Field opened.

The CEGB for whom the building was being constructed considered that electricity would be available, but were open to suggestion as to the fuel to be used. A study based upon night-time heat up, without normal day-time boost, indicated that the provision of an electro-thermal storage system would not increase significantly the instantaneous maximum demand.

There were no agreed tariffs by which financial viability could be assessed; if there had been there could be no surety that financial viability would not be undermined or negated in the future by changes in fuel prices and differentials.

Technically, the three methods under consideration were all suitable for integration with the proposed heat reclamation and re-cycling techniques.

It was decided that selection should be based upon:

• initial capital cost
• probable future availability of fuel
• suitability of the fuel for the purpose

If oil were used, the requirements relating to the chimney under the provisions of the Clean Air Act would have been incompatible with the overall building concept. The decision was taken to incorporate a gas fired low temperature hot water plant. However, the designs permit its removal and replacement by an electro-thermal storage plant quickly and conveniently should this become desirable in the future.

After studies incorporating various numbers and combinations, it was decided that seven modular boiler-burner units be provided. They are of high-low-off type, fully automatic in operation.

With seven units, it is possible to match the variable heating load closely, providing a higher average seasonal operating efficiency than would have obtained with fewer, larger units.

Table 5.9 summarises some of the data arising out of the studies, which was considered in the selection of the heating plant.

Table 5.9   Gas boilers – energy analysis

Type	Welded steel shell. Pressure jet burner + gas booster	Welded steel modular. Atmospheric burners. Natural draught	Welded steel modular. Atmospheric burners. Natural draught	Cast iron modular. Atmospheric burners. Natural draught
Total output	1 170 kW	1 125 kW	1 025 kW	1 050 kW
No of units	2	5	7	14
Control	Hi/Lo/Off	On/Off	Hi/Lo/Off	On/Off
Increments of total capacity control	4	5	14	14
Gas input	1 465 kW	1 406 kW	1 314 kW	1 315 kW
Electricity input	2.2 kW	0.05 kW	0.05 kW	0.05 kW
			Selected plant	

Control system

If the full potential of energy conservation designed into the system is to be realised, a suitable and comprehensive system of automatic control is essential.

The general principles of the control system are:
(1) Each air handling unit is controlled primarily through sensors that respond to the conditions in the area served by the unit.
(2) When the heating battery in any unit is on load, the cooling battery is off unless required for humidity control.
(3) When the cooling battery in any unit is on load, the heating battery is off unless required for humidity control.
(4) When the introduction of fresh air into any unit increases the heating or cooling plant load for the unit, the fresh air remains set at minimum.
(5) When an increase in the fresh air into any unit reduces the cooling plant load for the unit, the proportions of fresh air and recirculation modulate as necessary to obtain maximum free cooling without over-running and creating an unnecessary heating load.

The individual air handling unit control systems are integrated and co-ordinated into the main plant control systems, so that:
(1) The main boiler plant operates only when refrigeration plant is providing maximum heat reclaim, the recreational pool heat store is exhausted and there is still a heating deficit. Then the boiler plant operates under its own automatic controls to make up the shortfall.

(2) When heat reclaim exceeds the heating demand, and the recreational pool heat store reaches maximum temperature, the heat reclaim from air handling unit exhausts is reduced progressively to total shut-off dependent upon demand. This reduces both the water chilling and reclaim loads and the refrigeration plant load reduces accordingly.

(3) When heat reclaim exceeds the heating demand, the heat store is fully charged, and heat reclaim on exhausts from air handling units is off, then the surplus heat is discharged through the reject heat condensers and the cooling towers. This is the only time that any part of the thermal energy within the building is deliberately exhausted without re-cycling.

The whole of the automatic control functions are integrated through a central control console which permits up to 1000 connections to remote parts of the system. At this central console environmental conditions and the operating modes of the mechanical services plant are monitored and switched.

The central console operates in conjunction with outstation panels mostly situated in the plantrooms. The outstation panels are wired direct to the plant items. A standard multicore telephone cable connects the outstation panels to the central control console.

Alarms are indicated as either critical or general by flashing lamps and an audible alarm. Temperature and humidity conditions throughout the building can be monitored. Remote start-stop of all plant is incorporated. Plants can be switched on and off through any one of four programmable time switches, one of which includes a variable start time optimised by prevailing weather conditions and building temperatures.

Other facilities included are an alarm and status printer, plant diagram projector with slides for each plant, a temperature trend recorder and an intercom system.

The centralised control system permits one operator to monitor the complete building services installations. Should an operational fault occur anywhere in the systems it immediately transmits alarms to the central console. The operator can then despatch the maintenance engineers to the correct location with minimal delay. This saves the maintenance engineers from continuously touring the plant rooms to check plant operation and status.

The status print-out and temperature trend recorder permit checking and re-setting of plants and controls to effect maximum energy conservation. The remote stop-start facilities permit variation in the combinations of air handling units running to match occupation and use of the building; this also contributes to maximum energy conservation.

Summation of design

Whilst the building does not provide the ultimate in energy conservation, a building that did might not have been as suitable because of other considerations. The brief emphasised the need to exploit the natural amenities and, at the same time, provide a building of sensitive design with conservative energy demands. The brief has been fulfilled well in relation to these parameters.

If an attempt were made to list the merits of the air conditioning design, they would include:

(1) The use of water circuits to transmit thermal energy from the central heating and cooling plants to the outlying air handling units, in combination with low velocity air ducts, requires less motive power than other systems.

(2) The use of the refrigeration plant to up-grade and re-cycle internal heat gains, provides a better re-utilisation factor than straightforward air recirculation. The water in the reclaim condenser circuit is a more suitable heat transfer medium, at a more suitable temperature. Its flexibility of operation and control is compatible with that of the decentralised air handling units. It promotes the effective transfer of heat from parts of the building where there is a surplus to those areas where a deficit exists.

(3) The multiple decentralised air handling units permit greater choice and flexibility when (a) the building is not approaching maximum heating or cooling loads and (b) when it is not fully occupied. They can be switched on and off from a central location to reduce electrical consumption.

(4) Whilst the possibility of a breakdown on air handling plant might be increased, the effects of such breakdown would be minimal.

(5) There is no encroachment upon usable floor space, particularly around the perimeter. No additional space was needed in ceiling voids.

(6) The system is quiet. All that the occupants see is the decorative, acoustic ceiling.

Air conditioning for telephones

The telephone system incorporates a fully computerised private branch exchange. This necessitates air conditioning in the apparatus room to standards adopted in computer rooms.

Because of the special requirements and the need for continuous operation of the air conditioning, a completely separate system is provided. It includes:

• a refrigeration machine of reciprocating compressor type with capacity control, a chilled water evaporator and a water-cooled double bundle condenser for heat reclamation

• an air handling unit with duplicate fans, for primary ventilation and environmental control.

The operating cycle for the plant is the same as that described for the main air-conditioning system, incorporating heat reclamation and free cooling. To optimise plant utilisation and energy conservation, the chilled water, hot water heating and heat rejection circuits are inter-connected with those of the main air-conditioning system.

HOT WATER SERVICES

Hot water for kitchen and toilets is provided by two storage calorifiers located in a basement plant room. The primary circulation to the calorifiers is connected to the gas-fired boiler plant.

A non-storage calorifier is installed to pre-heat the cold feed. It is connected on the primary side to the reclaim heating distribution network. The automatic controls are linked into the central system so that reclaimed heat is not rejected through the cooling towers when it can be used to preheat the domestic hot water supply cold feed.

As a further energy economy, manual mixing minispray taps are used in toilet areas.

THE ANNUAL ENERGY BALANCE

The effects of alternative lighting intensities

The CEGB brief for the open plan office areas required that the lighting should provide an even overall illumination level together with an adequate 'sideways' component, and that 500 lux should be considered in conjunction with tinted glass.

At the time the designs were initiated, the intensity of artificial lighting recommended by the IES for open plan offices was 750 lux. A large number of buildings had been constructed during previous years with an intensity of 1000 lux or more.

It might be considered that high intensities can be justified by use of modern techniques of heat recovery and resultant low thermal balance points, since:
- the higher capital costs of the lighting system would be offset, to some extent, by the relatively low cost of small capacity electrothermal systems for supplementary heating, and
- running costs for supplementary heating would be negligible in a building which is virtually 'self-heating'.

Whilst such submission might be credible, it is not necessarily viable. Actual running costs of such buildings did not appear to confirm the principles.

Accordingly, these design principles were being criticised. A contrary opinion was common that lower intensities were desirable, despite higher thermal balance points. The advantage could include:
- lower capital and running costs of the reduced capacity electric lighting system
- lower capital and running costs of the air conditioning system because of reduced cooling loads due to lighting
- the two preceding factors would offset increases in capital and running costs of supplementary heating.

The suggestion in the CEGB brief that an intensity of 500 lux might be adequate in conjunction with tinted glass emphasised the need to analyse the alternatives.

A preliminary study indicated that the capital cost for a 500 lux intensity solution would be in the order of 20% less than that for a 750 lux intensity design. It was apparent that annual average energy requirements must be a paramount factor in the design decision, both from aspects of cost-in-use and energy conservation.

It was decided to make further studies, based upon the total average annual energy requirements for lighting, cooling and heating. Since tinted glass was being used, with internal venetian blinds on sunny elevations to reduce glare, it was decided that the studies should be based upon intensities of 500, 750 and 1000 lux.

Daylight studies established that a 2% daylight factor produced an illumination level of approximately 100 lux at about 4m from the perimeter, rising to approximately 10% at about 1.5m, providing about 500 lux on the working plane. The studies were based on outdoor illuminance of 5000 lux

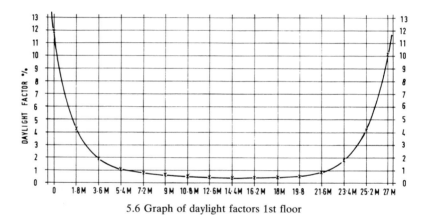

5.6 Graph of daylight factors 1st floor

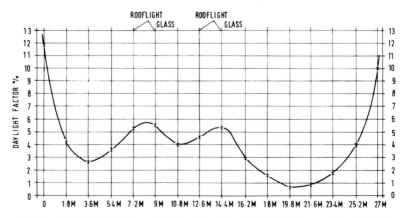

5.7 Graph of daylight factors 2nd floor (combined effect windows and roof light)

with an overcast sky (see figs 5.6 and 5.7). It was apparent that minimum daylight levels would be exceeded for many hours per annum and that automatic switching around the perimeter and under the roof lights, related to daylight characteristics, would reduce energy consumption appreciably.

Studies encompassing maximum instantaneous and annual average energy consumption were undertaken covering illumination values of 500, 750 and 1000 lux and automatic switching. Fig 5.8 is a typical heat balance for instantaneous loads. Fig 5.9 is a heat balance for annual average loads with lighting levels of 500, 750 and 1000 lux.

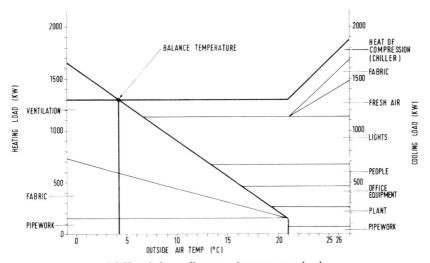

5.8 Heat balance diagram – instantaneous loads

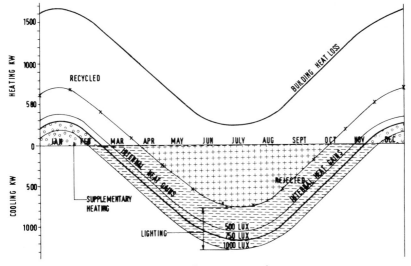

5.9 Heat balance diagram – annual average

The analyses provided the following data.

	Lux		
	500	**750**	**1000**
Maximum instantaneous lighting load	399 kW	554 kW	706 kW
Maximum instantaneous cooling load	1547 kW	1702 kW	1854 kW
Maximum instantaneous heating load	1140 kW	1140 kW	1140 kW
Maximum instantaneous heat recovery	531 kW	719 kW	880 kW
Thermal balance point	11°C	7°C	4°C
Average annual energy consumption	2588 MWh	2786 MWh	2850 MWh

The study confirmed expectation that a lower lighting intensity entailed:
• a low maximum cooling load
• a lower maximum heat recovery rate
• a higher supplementary heating load.

It is also demonstrated that whilst the lower lighting level provides less reclaimed heat
• a greater proportion of it can be used for a greater proportion of the year
• the overall energy equation is improved, and
• the average annual energy consumption is reduced.

The studies confirmed that for optimum energy conservation the intensity of artificial lighting should be restricted to that necessary for adequate illumination. Lighting levels should not be up-graded nor justified on the grounds of increased heat gains available to heat recovery systems and lower thermal balance points.

The purpose of these high levels of illumination was therefore examined to determine whether illuminances of this magnitude were necessary for deep plan designs.

Many articles have been written on the subject and the relevant observations noted.

An inspection was made of a number of buildings with varying lighting levels and window glazing ranging from clear pane to glass with very low light transmission factors with varying colour tints.

The inspections and discussions with users indicated the following major reasons why lighting levels in excess of 500 lux, which is the recommendation for small plan offices, were deemed necessary.

(1) To reduce the contrast between perimeter brightness and internal office brightness, ie 'visual discomfort' caused by glare. This was particularly evident where glass of a high light transmission had been used, even with high artificial lighting levels.

(2) To prevent an atmosphere of gloominess in the deep plan view which might prove distressing for the occupants, even though illumination for the localised task would be adequate.

(3) To ensure that the ceiling was adequately lit to give the interior a feeling of spaciousness in height and in particular minimising the long view tunnel effect.

(4) To incorporate an adequate sideways component to the lighting to illuminate satisfactorily vertical surfaces and to provide good modelling to personnel to ease the visual task.

The various aspects of achieving a satisfactory lighting design, as referred to, were considered in detail. Arising out of these considerations, it was decided that a satisfactory internal environment could be achieved with a lighting intensity of 500 lux provided certain design concepts could be implemented. These are:

(1) The selection of a perimeter glass providing about 30% light transmission to prevent undue perimeter glare, whilst retaining the external viewing facility and appreciation of outside conditions. The glass should have a natural colour rendering if possible, preferably in the gold range, rather than green or grey. Glasses incorporating these characteristics were available.

(2) Many lighting installations in deep plan buildings with high illumination levels have employed a fully recessed low brightness fitting with vertical lighting characteristics. This may have been desirable from the aspect of glare but it often results in a ceiling that is poorly lit with vertical surfaces at higher level in shadow in contrast with the highly lit working surfaces. This has the effect of lowering the ceiling height and providing poor modelling. To eliminate these difficulties it was decided to design a special lighting fitting to light the ceiling and also provide a sideways distribution element. Such a fitting would light the vertical surfaces to the highest possible level without compromising the glare zone.

(3) The lighting fitting should have sparkle without undue brightness to bring the ceiling zone to life. It was decided that this could be achieved by the specially designed lighting fitting, with semi-recessed reflector and louvre assembly, in conjunction with a shaped ceiling containing a variety of angled surfaces.

(4) Even with such provisions an illumination level of 500 lux might create an air of gloominess when vision was focussed to very distant areas of the deep plan; the absence of brightness from the lighting might cause this. To overcome this, it was decided to deliberately attract vision to shorter distances by providing highly lit vertical surfaces or objects of interest, ie planters, murals and the like; this would have the effect of taking the deep plan view out of the large interior.

It was considered, therefore, that providing a suitable integration of ceiling and lighting could be achieved, and that walls and floor finishes had a sufficiently high reflection factor, then a level of 500 lux could provide a satisfactory working environment.

Following the design studies a decision had to be taken on the illuminance level to be used as a design criterion in relation to energy conservation. Whilst there were indications that 500 lux would be optically satisfactory, there were no similar buildings with such a low intensity that could be used for comparative assessment. Also, at that time it was not possible to

DATA SUMMARY

Building volume	57 536 m³	
Floor area	19 600 m²	
Max cooling load	1 680 kW	
„ Heating load	1 600 kW	
Capacity of refrigeration plant	On reclaim cycle	On reject cycle
(a) Evaporators		
Flow rate	65.86 l/s	65.86 l/s
Temp differential	12°C to 5.5°C	12°C to 5.5°C
Cooling capacity	1 742 kW	1 742 kW
(b) Reclaim condensers		
Flow rate	48.0 l/s	–
Temp differential	40°C to 30°C	–
Heat reclaimed	2 008 kW	–
(c) Reject condensers		
Flow rate	–	48.0 l/s
Temp differential	–	38°C to 27°C
Heat rejected	–	2 210 kW
(d) Motive power input	490 kW	445 kW
Capacity of boiler plant		1 025 kW
Heat storage capacity of pool		
Nominal water quantity		135 000 litres
Max storage temp		30°C
Min storage temp		25°C
Nominal heat storage		3.40 GJ
Total number of air handling units		33
Total supply volumetric flow		116 m³/s
Average air change rate		7.29 per hour
Thermal balance temperature		4°C
Maximum electricity demand		1.3 MW
Installed lighting load		450 kW

Cooling load analysis

Windows	8%
Walls	2%
Roof (including roof lights)	2%
Floor	0%
Fresh air	22%
Lighting	26%
Office equipment	11%
People	14%
Motive power for mechanical services	12%
Sundry	3%
	100%

Heat load analysis

Windows	14%
Walls	4%
Roof (including roof lights)	13%
Floor	1%
Fresh air	63% *
Sundry	5%
	100%
Heat reclaim	61%
Boiler load	39%

* There are large areas on ground floor which operate on 100% fresh air at maximum load, eg kitchen, restaurant, lecture theatre, meeting rooms, reprographic etc.

determine exact locations within the building where higher levels from the ceiling system might be required for drawing office tasks. It was decided, therefore, that the electric lighting installation should be designed to provide an illuminance of 750 lux.

It is interesting to compare the design study for 750 lux with the summary of final design data at the end of this chapter. Meticulous attention to energy conservation during the final design of the lighting installations effected a reduction in installed capacity from 554 kW to 450 kW whilst maintaining the prescribed illuminance.

It was decided that ventilated luminaires be used. Apart from any other advantages, the removal at source of a large proportion of the heat generated permits a greater temperature difference between supply and return air without adversely affecting the room conditions. This reduces the mass flow through the air handling units and, consequently, the energy consumed by the plant.

Design could now proceed on the basis of:

- an illuminance of 750 lux in deep plan areas and 500 lux in small partitioned offices
- solar switching at perimeter and under roof lights
- ventilated luminaires.

For control purposes each open office floor is divided into four, with a manual control unit in each service core. Each control unit provides the following switching functions:

- one-half of the general office lighting equally distributed over the floor area, supplemented by automatic solar control
- full general office lighting, supplemented by further automatic solar control
- essential lighting area 1
- essential lighting area 2.

The half lighting is for use when the building is occupied by cleaners etc.

Each of the main switching functions controls the operation of a number of sub-circuits, each controlled by miniature circuit breakers. By the operation of selected miniature circuit breakers local control of lighting is achieved; for example, a limited area only could be fully lit for a small group of people who may be working when the rest of the building is unoccupied, thereby effecting energy saving.

BIBLIOGRAPHY

Electricity Council, *Integrated Design – A case history*, September 1969, p 20.
N. O. Millbank, J. P. Dowdell and A. Slater, *Investigation of Maintenance and Energy Costs in Office Buildings*, Building Research Establishment Current Paper CP38/71, JIHVE Vol 39, Oct 1971, pp 145–154.
A. F. C. Sherratt (Ed), *Integrated Environment in Building Design*, Barking, Applied Science Publishers, 1974.

A. F. C. Sherratt (Ed), *Energy Conservation and Energy Management in Buildings,* Barking, Applied Science Publisher's, 1976.
A. G. Louden, *Summertime Temperatures in Buildings without Air Conditioning,* BRS CP 47/68, Building Research Establishment, Garston, 1968.
A. C. Hardy and P. E. O'Sullivan, *Insulation and Fenestration,* Stocksfield, Oriel, 1967.
BSI, *Code of Basic Data for Design of Buildings,* Chapter 1, part 2, artificial lighting.
CIBS, *IES Code for Interior Lighting,* Chartered Institution of Building Services, London, 1977.
CIBS, *Depreciation and Maintenance of Interior Lighting,* Chartered Institution of Building Services, London, Technical Report No 9, 1977.
CIBS, *Daytime Lighting in Buildings,* Chartered Institution of Building Services, London, Technical Report No 4, July 1972, Supplement 1977.
BRE, *Estimating Daylight in Buildings 1,* BRE Digest 41, Building Research Establishment, Garston.
BRE *Estimating Daylight in Buildings 2,* BRE Digest 42, Building Research Establishment, Garston.
Billington N.S., *Air Movement over Hot or Cold Surfaces,* Laboratory Report No 29, BSRIA, Bracknell, 1966.

DISCUSSION

W. J. Crowther (Department of Finance, Belfast)   The importance of air leakage from both new and existing buildings has been stressed. I would like to ask how is it proposed to measure air leakage? And what effect would such measurements have on the assumptions made when applying design formulae to the sizing of air-conditioning plant?

T. Smith (Steensen Varming Mulcahy & Partners)   It should not be necessary to measure air leakage as it should be eliminated. Where leakage does occur I do not know how it would be measured for a whole building. Attempts can be made to avoid it by good standards of construction, not only of window frames which traditionally leak, but brickwork, topwork, doors, etc. Cold bridges are also important where heat is transferred by conduction to the exterior. It is quite remarkable how many buildings have areas that could not comply with the basic fundamentals of good insulation. Designing and installing closely controlled environmental installations in an envelope leaking both heat and air is just not common sense.

P. Day (City of London Real Property Co Ltd)   On very tall tower blocks wind speeds of up to 27 m/s (approx 60 mph) have been recorded for a number of hours.

A British Standards Institution Draft* for Development DD4 deals with

*BSI Draft for Development, Recommendations for the Grading of Windows: Resistance to wind loads, air infiltration and water penetration and with notes on window security, DD4: 1971.

grading of windows with respect to resistance to wind loads, air infiltration and water penetration. I am of the opinion this draft does not set a sufficiently high standard for air-conditioned buildings, particularly considering energy conservation.

The architect and consultant should prepare their own specification to meet the air infiltration allowance set by the air-conditioning engineer. For refurbished buildings it is common to find distorted window frames and if air conditioning is being installed a large percentage can be sealed with a mastic and screwed up. The windows that must be openable to meet statutory regulations may have to be replaced to keep the rate of infiltration to acceptable levels.

Problems with air and water leakage also apply to curtain walling where it is essential proper sealing between panels is carried out and inspected by the Clerk of Works. On very large projects a full scale modular test of a typical panel, window frame and glazing using a wind tunnel and sprays may prove extremely valuable.

R. F. Wallace (Colt Solar Control Ltd) Differences in emphasis as to the contribution of solar heat gain to cooling loads, have been given by different authors. Sir Alan Pullinger (chapter 1) suggested it was not particularly significant, whereas Mr Allford (chapter 4) indicated it was a highly significant element, going into detail on the benefits of external shading. Reflective glass is suggested as a possible alternative although reflective glass allows a certain amount of direct radiation to enter. There has been discussion recently not only on the effects of direct radiation but also the long wave radiation emitted from tinted glasses, because of the high surface temperatures they can reach which can be of the order of 62°C. However, how much of this solar heat actually contributes to cooling loads? There will obviously be wide variation due to the amount of glass, the location, the depth of the building etc, but can anyone provide recent examples? A comparison made about four years ago between double clear glass and internal blinds and double clear glass and external blinds indicated a 17% reduction in the capital cost of the air-conditioning plant and the blinds, 32% reduction in the energy costs.

J. S. Torrence (Steensen Varming Mulcahy & Partners) An example Mr Wallace did not mention but which was referred to in the paper is the use of solar resistant glass, a fixed internal Venetian blind and a further sheet of single glazed clear glass window/wall with room air exhausted between the two sheets of glass, thus reducing the load by about 'one-third'. I am not suggesting that that is an energy saving solution, but rather a way to stop extreme solar gain in lieu of fitting external shading.

B. Rimmer (Loughborough University of Technology) I wonder if Mr Smith could elaborate on this window extraction unit. Obviously the heat into the room is largely radiative so one wonders how drawing air through two panes of glass will pick up that heat and control gains to the extent suggested.

T. Smith The scheme for which the ventilated window was designed, developed and used was the responsibility of my partner, Mr Jack Torrence, and is I believe the only one in this country. The design is intended to maintain a controlled inner glass temperature at a level such that people are comfortable sitting very close to windows. It saves resources by allowing the maximum number of people to be accommodated for a given floor area. The air passing between the two sheets of glass picks up, by convection, the gains or the losses on the outer skin. The air sandwich formed maintains the inner skin temperature close to the room temperature.

L. J. Wild (George Wimpey & Co Ltd) At present we have under test at BRE a combined pre-cast panel and window, sealed with mastic. I agree with Mr Day that such a test is the only way to check the sealing of windows and cladding for leakage. It is expensive so it would not be possible to afford tests in every instance.

D. Arnold (Troup Bywaters & Anders) Mr Day (chapter 6) gives an analysis of about 11 buildings and the average cooling load is approximately 0.22 kW/m². Quite surprisingly the load on the building described by Mr Smith is about half that, less than 0.1 kW/m². Obviously the designers of that building have taken several measures to reduce the air-conditioning load and I wonder what positive steps were taken with that in mind and what else they would have liked to do.

V. A. Hammond (Steensen Varming Mulcahy & Partners) In the compilation of the design, a large number of energy targets were established for many elements contributing to the total energy need. No average load related to floor area was stipulated. No comparison was made with other buildings. Each individual element of the building which directly affected thermal load was studied in detail and the (often conflicting) parameters optimised. Evidence of the success of this approach is inherent in the question. We have gone into some detail about the various steps taken to economically reduce load and the final load is a complex interaction of all of them.

H. A. Rudgard (Shell UK Admin. Services) In his presentation Mr Smith used the quotation 'small is beautiful' which I assumed to imply a recommendation to use a multiplicity of small plants: a recommendation which results in a larger space required. On the other hand Mr Day (chapter 6) said that office space was at a premium and plant room space should be minimised. As one of the engineers responsible for running and maintaining building services plant I prefer larger plants with adequate space for maintenance. A compromise is obviously necessary but could be difficult to strike.

P. Day Maintenance and plant space are obviously important. If you are designing a building with yourself as the ultimate client, the situation is slightly different from the developer who must be interested in maximising floor space to rent.

There are ways of minimising plant room space and I can give a specific example. For the system shown in chapter 6, fig 6.4 the access to the individual floor air handling units was arranged via large double plant room doors. Access to filters, coils, fans, controls etc is relatively easy and there is no need to allow any more than 75 mm between the unit and door. For roof-top plant rooms access should be provided through the roof and walls. In the case of walls hinged louvres are very practical with removable planks at roof level.

The most important factor is to consider the maintenance and removal of plant and equipment at the design/planning stage.

T. Smith Mr Day is not advocating minimising plant room space, to the extent that it is unmanageable and incapable of maintenance and operation. Neither am I advocating maximum space. It is a compromise we are all seeking, not too much and not too little. The realisation as a young engineer on his first visit to the USA of the differences in the American and the European attitudes on space provision for services still lingers with me. An American engineer needs to take a pipe or duct from one point to another and that is exactly what he does, but in the UK it goes twice up and down over light fittings, bends over some other piece of obstruction etc in the trivial little ceiling space allocated to him. What must this be costing both in his effort and the continuing energy cost to his client for pressure losses? As I have said, through the whole life span of the building he will pay a heavy premium for energy. Far better to increase the building height by 5%. This would not cost 5% of the building cost, but it can save a great deal of money on services, both capital and revenue. I am highly critical of the restrictions imposed upon services designers in the UK primarily by some unenlightened architects.

Dr I. W. Weller (Adams Green & Partners) I fear we might be misled by the term 'energy optimisation' whereas we are generally all concerned with energy minimisation.

In an earlier response Mr Smith described how air is drawn across glazing, thereby keeping the glazing surface temperature very close to the air temperature. I agree this would improve comfort conditions, but at the expense of energy control, simply because the glazing is kept at the air temperature which in the winter time will increase heat loss through the glazing and during the summer time will increase the heat gain to the ventilation systems or worse still to the air-conditioning systems.

T. Smith In fact, the reference related to another building. We did not claim that the subject building is a minimum energy building but rather an 'energy conscious' building. The success of the approach is emphasised by the low average annual energy imput. This was made clear in preparatory discussion. I am sorry if you have been misled but we did not set out to mislead you and I do not think we have.

J. S. Torrance Regarding glazing, do not forget the building was designed in 1970, well before the energy crunch. The method of extracting air between the two sheets of glass was used because it was a way of dealing with

a large expanse of glass to produce what is I think a very nice glass building.
A. S. Kennedy (Property Services Agency)   If we accept that there is a specific need to air condition a building and we have been sensible about plan form, orientation, fenestration, thermal transmittance etc it seems there is still another area worth working at – the level of illumination within the building. I would be grateful to hear of any experience of successful application of task lighting coupled with a lower level of general lighting as an energy conservation measure.
T. Smith   We have not been successful in getting any clients of any size to opt for this totally. Basil Gillinson was the architect on the CEGB Building and we worked closely together on lighting. Perhaps he would say something about the lighting of the building and the resistance we met. Our own Dublin office uses task lighting – it was a tram depot we converted into a mini-Burolandschaft office. The designs there are based on low-background with high task lighting.
B. Z. Gillinson (Gillinson Barnett & Partners)   When we designed the CEGB headquarters at Harrogate, we did look into all aspects of lighting design. Our clients, being very much in the electrical industry, were very keen that every possible approach should be investigated. We examined the experimental areas at Pilkingtons, St Helens, where the principles of minimum background lighting, combined with maximum task lights are being tested and felt, at the time, that there was insufficient evidence to show the arrangement would be entirely satisfactory. The main problem we saw when designing a large office building, was that our clients were looking for long term flexibility with potential changes in the use of space for differing functions. We were concerned with our consultants to tie in the design for flexibility with the mechanical systems so that in the long term we could achieve a sensible balance. The use of task lighting could lead to problems, if for example at some time a high density of use is introduced into a particular space with subsequent provision of very high task lighting levels, eg in a large drawing office. Such an intensive use of a particular area would introduce problems by imposing additional loads on the air handling plant. Should energy become enormously expensive in the future, we may well arrive at, or perhaps more properly, return to a situation where each person has his own individual lighting source. At the present time we think that the right approach is to provide a minimum acceptable lighting level, which we consider to be 500–600 lux.
E. C. Lovelock (Shell UK Admin Services)   I visited a Government office in Stockholm known as 'Garnisonen' which was completed about 1970. It has a gross floor area of something like 150,000 m², part of which is for cellular type offices and part landscaped. The interior design is a very brutal form, with all services eg wiring, pipes, ductwork etc left exposed and the lighting installation is designed for task lighting. The background or a main overhead lighting system provides a cut-off plane just below all the exposed services. This fluorescent lighting system is supplemented by a second

electrical track from which tails may be dropped to desk lights or spotlights as required.

P. G. T. Owens (Pilkington Brothers Ltd)   The task lighting experiment carried out in our offices three or four years ago is fully documented in several places*. So far as modern buildings are concerned a good example of the use of task lighting is the CEGB building at Beaminster Down, near Bristol. This building is fully task lit, and like the Harrogate building other energy conservation features are included.

A. Conlon (Varming Mulcahy Reilly Assoc, Dublin)   Our own offices described by Tom Smith were designed for a lighting level of 300 lux with the intention of having task lights for draughtsmen. When we actually went into the building we found that many of the draughtsmen were quite happy with 300 lux and very few asked for the task lights that were available.

*P. G. T. Owens, *Energy conservation and office lighting,* Proceedings of the International CIB Symposium on Energy Conservation in the Built Environment, 6–8 April 1976, Building Research Establishment, Garston.
C. Cuttle and A. I. Slater, A Low Energy Approach to Office Lighting, *Light and Lighting,* 1975 (68), pp 20–24.
C. Cuttle and A. I. Slater, The Effective Use of Energy for Office Lighting, presented to CIE Conference, Session TC4.1, London, 1975.

6 Refurbished buildings and air conditioning

P. H. DAY

INTRODUCTION

During the past few years the construction of new commercial office building has been very limited, some suggested reasons being government controls, taxation, very high construction costs, over supply of office space and low rental levels.

Developers have turned to the refurbishment of existing buildings but in many cases air conditioning has not been included in the programme of works. There are a number of reasons for this, such as:

- high capital cost – approximately four to five times that of central heating
- extended construction period
- difficulty in recovering the costs by higher rental charges
- high operating costs especially due to the recent rise in basic electrical and fuel charges. Fig 6.1 shows the refrigeration load/floor area for eleven buildings.
- many of the buildings being refurbished were originally constructed in the 19th century with very thick walls and small areas of glazing.

Financial considerations have a very marked influence on the construction industry and it is extremely difficult to make a case for air conditioning in the present economic climate. A further problem associated with finance is the standard of air conditioning installed. In the more speculative area of commercial development the performance of the low capital cost air conditioning systems installed have left much to be desired and caused the industry to be the subject of some adverse comment.

It is perhaps a little early to predict a return to the early 1970 period when air conditioning was nearly always installed. If this does happen refurbished buildings will be a major market for the industry. In the interim there are some refurbished buildings where air conditioning is needed now and these will be examined in this paper.

REFURBISHED BUILDINGS WHERE AIR CONDITIONING COULD BE CONSIDERED

Over recent years there has been considerable debate as to the wisdom of constructing the lightweight high rise buildings of the 1950s with their large

108

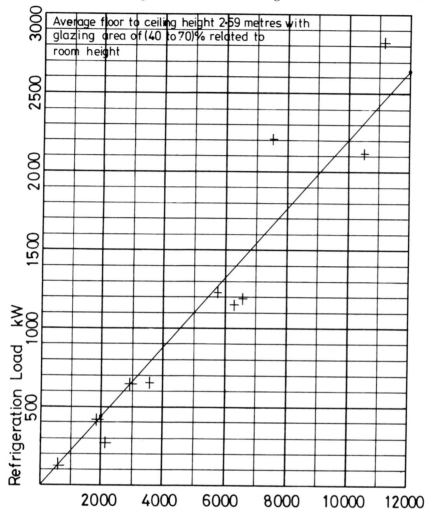

6.1 Refrigeration load plotted against floor area available to let
  Graph based on following data
1 External conditions
    Maximum summer 30°C Dry bulb   Minimum winter −1°C Dry bulb
                    20°C Wet bulb   100% Saturation
2 Occupancy levels
  1 person per 5.50 square metres
  Fresh air rate/person 11.80 litres/second
3 Lighting intensity 30 watts/square metre
4 Machine load 10 watts/square metre
5 Internal conditions Dry bulb 20°C to 22°C
                    % Saturation 38% to 50%
  Temperature swing to 24°c during peak heat gain for 2 hours
6 Correction factors for refrigeration loads
  (a) Occupancy x 0.85 (b) Lighting x 0.95 (c) Office machines x 0.9
7 Lightweight building – no blinds but Solar Control glass
8 Data based on 11 buildings

6.2 Typical Victorian building

glass window areas and curtain walling. The high heat losses and gains indicate an era when fuel was cheap and air conditioning very rarely installed, but air conditioning could now be considered for buildings like this if refurbished.

Since 1965, in spite of past experience, buildings with large glass areas have been constructed and some of the air conditioning plant in them is ready for replacement and refurbishment.

Fig 6.2 illustrates a typical Victorian building which looks very attractive after stone cleaning and general repairs have been carried out. Note the heavy construction and size of windows, a marked difference to the modern era of architecture. Simple heating would usually be adequate for a satisfactory environment because of the heat storage effects of the mass in the building fabric. There is, of course, no suggestion that air conditioning should not be installed in heavily constructed buildings since better environmental conditions will result where a satisfactory case can be made for the additional capital expenditure needed.

When the building to be refurbished has not been listed by the authorities as of architectural merit developers are in some cases minimising expenditure, anticipating that when market conditions are favourable it will be demolished, and replaced by a building of modern construction.

Refurbished buildings which must be considered are therefore those whose construction demands air conditioning or where the present air conditioning equipment requires replacing. A further consideration would be improvement in environmental conditions, such as reduction of traffic noise.

PROBLEMS ASSOCIATED WITH REFURBISHED BUILDINGS

There are considerably more problems associated with existing buildings than those of new construction, especially when air conditioning is being considered and these are discussed in the following paragraphs.

Space

The developer will probably be keen to provide as little room for services as possible since maximum space means a higher rental. Adequate space for services is essential. The architect and services designer must try to be as firm as possible about this since otherwise the overall result could well turn out to be disastrous. However, the skill of the designer should be used to occupy the minimum of space rather than produce the 'ball rooms' of plant area seen in a number of buildings. A reasonable initial figure to use for all services is 8 to 12% of the building volume.

Care must be exercised in surveying an existing building. It is very common to find floor slabs not level. The relationship between the floor and soffit of the floor above may not be parallel with in some cases a divergence or convergence. Some post-war modern buildings were built in the days of speculative office development with minimum height between floors. It is not uncommon to find a floor to soffit dimension of 2.718 m. The engineer will then have to design services within a false ceiling void space of 120 mm. With very careful planning and considerable attention to detail this is possible and the designer must not over economise on drawing office time. The extra work involved may present a difficulty especially with the fee structure for services. It generally costs more to design air conditioning properly for a refurbished building than for new construction.

Building structure

For reinforced concrete structures it is not always possible to provide holes in the most convenient places and this can cause major problems. A vertical perimeter distribution system for a perimeter induction system may be selected by the services engineer only to find that after initial agreement by the structural engineer a few pilot holes in the appropriate places show the presence of reinforcement.

There are several possible reasons for unexpected presence of reinforcement – for example, less care may have been exercised in keeping detailed drawings, supervision on site may not have been to the same high standards

6.3 Existing heating installation distorted bellows

as today and the pressure for rapid construction in a speculative environment. Experience suggests that it is far better to design services with the minimum of disturbance to the existing structure. Even when record drawings are in existence they are not always accurate. It is very embarrassing for the professional team to firm up a scheme, agree it with the client and then find major problems during the construction period.

Existing services
Very often the existing heating system can be retained and revised as part of the air conditioning, although a careful survey is needed. An illustration is given in Fig 6.3 showing a distorted bellows which formed part of an existing low temperature hot water heating installation in a building due for refurbishment. Expert opinion suggested that the distortion was due to over pressurisation, but the photograph shows a small wedge shaped piece of wood between the bellows and the building structure. It is still a matter of speculation why this wedge was used.

The need to carry out a detailed survey (with record drawings, if necessary) of the existing services especially if the installation is over ten years old cannot be overemphasised.

Programme
With a refurbished building the major cost will probably be for the services, 50 to 60% of the total cost not being uncommon.

It is very important that the services contractor is appointed very early so that his experience can be used to formulate an overall programme in association with the main contractor.

This could mean that a negotiated contract is the best method but if competitive tenders are called for the services consultant must complete the majority of his design and drawing office work earlier than would normally be the case for a new building. It will also be necessary to call for a services installation programme as part of the tender submission. The design team will have to show installation activities for the main contractor's work which should not be difficult since in most cases very little major construction is called for. The most important factor will probably be speed since with the structure existing the client will require to recover such costs as compensation to tenants, construction, loss of rent during construction, advertising, fees etc as quickly as possible.

Specification

It is very important to be realistic when considering the content of a specification. It is now common practice for a prospective tenant to obtain a report from a consultant before a lease is agreed.

It is preferable if basic information is provided in a small document under the following headings:

- internal and external environmental conditions
- occupancy rates and fresh air requirements
- machine loads
- design data
- lighting levels
- requirements of the local authority
- type of air-conditioning system
- operation of air-conditioning system

This information should be helpful to all professional advisers.

General

With refurbishment work there is always the temptation to provide more finance for building finishes and skimp on the services. A compromise is generally reached and the specification given below is considered reasonable for the London area.

Specification for the London area

1. Internal and external conditions

Internal: Maximum summer 22°C and 50% saturation allowing the temperature to swing to 24°C for two hours at peak heat gain.
Minimum winter 20°C 38% saturation.

The lower percentage saturation can result in considerable fuel savings. Experience gained in two buildings where no winter humidification was installed suggests that values of 35 to 38% saturation give rise to very few complaints.

External: Summer   30°C dry bulb
                                 20°C wet bulb
              Winter   −1.0°C dry bulb
                                 100% saturation

The sling wet bulb temperature is the most important temperature to consider since it effects the size of refrigeration plant chosen and other plant such as pumps, cooling towers, etc.

Measurement of sling wet bulb temperature taken in the City of London over the past few years suggests that the large water mass of the River Thames elevates the wet bulb temperature above the level recommended for design. Near the river in London a realistic value is 20°C.

2. Occupancy rates and fresh air requirements

Sections 5 and 7 of the Offices, Shops and Railways Premises Act 1963 deal with overcrowding and ventilation.

Unfortunately section 7 is rather vague and the general guide issued by HMSO gives very little help.

The best solution is to agree the occupancy level and fresh air rates with the Public Heath Office of the local authority.

Discussions with the Corporation of London suggest minimum fresh air rates per person of 0.0094 m³/s but recommend 0.0118 m³/s. They further recommend that the density of occupancy should be related to approximately 11.5 m³ per person the limit under section 5 of the Act where ceiling heights are below 3.050 m.

With ceiling heights in order of 2.59 m the density of occupation could be as high as one person per 4.40 m².

Floors occupied by executives and senior management generally have a very low level of occupancy but on the other hand general administration departments employing junior staff can give rise to maximum occupancy as defined in the Act.

To supply more fresh air than is required will result in higher operational costs and the system designer should try and introduce some flexibility into the design to deal with this problem, an aspect of design discussed in more detail later.

A reasonably specification would be:

1 person per 5.5 m² with a fresh air rate of 0.0118 m³/s per person, with a flexibility in the design to change these values to meet actual loads.

3. Machine loads

An allowance of 10 watts/m² will give reasonable cover for typing machines, small desk top calculators etc.

4. Design data

This section should cover important items such as authority for design data ie CIBS/IHVE Guide, if solar loads are calculated on a light or heavy weight building basis, air velocities in the air-conditioned space, pressure drops for selecting pipe sizes and air ducts. The items mentioned are not intended to be a complete list but a guide for consideration.

5. Lighting levels

Experience suggests that lighting levels above 500 lux are unnecessary. A general lighting level of 350 to 400 lux supplemented by a desk lamp with a 60 watt tungsten bulb for close work has been found to be very satisfactory. High levels of illumination are gross users of energy even when reclaim techniques are used and should be avoided other than for special areas.

6. Requirements of the local authority

Within the GLC area a formal consent containing a large number of conditions is normally provided for each building. These consents now contain specific requirements for the proper maintenance and inspection of plant and services and also the employment of competent persons. If a full repairing and insuring lease is proposed then this information must be passed on to the tenant since the consent condition will become a statutory responsibility.

7. Type of air-conditioning system

A brief description and diagrammatic drawing of the system together with a list of advantages for the particular equipment and plant used.

Experience has shown that information of this kind is extremely helpful to architects, surveyors and agents.

8. Operation of air-conditioning system

Details of the method of operation, for example, fully automatic or requiring staff to operate the plant. This information is extremely important since the cost of maintenance staff can be as high as £10,000 per annum/person when basic salary, overtime, pension payments and general overheads are taken into account.

TYPE OF AIR-CONDITIONING SYSTEM

This will depend on the brief given by the client. If open plan offices are required then lower cost and operational levels will be possible when compared to individual offices.

Over the last few years enquiries to tenants and agents indicate that for the City of London the requirement is generally for individual offices and flexibility to allow easy movement of partitions. Obviously the design must suit market conditions, even if first cost is high. The additional cost would be in the order of 20 to 25% higher than open plan.

The main objectives for refurbished buildings are to disturb the existing structure as little as possible, speed of installation, and recovery of development costs.

Four pipe fan coil system

An air-conditioning system which in practice has been found to meet these criteria is one utilising four pipe damper controlled fan coil units. Fig 6.4

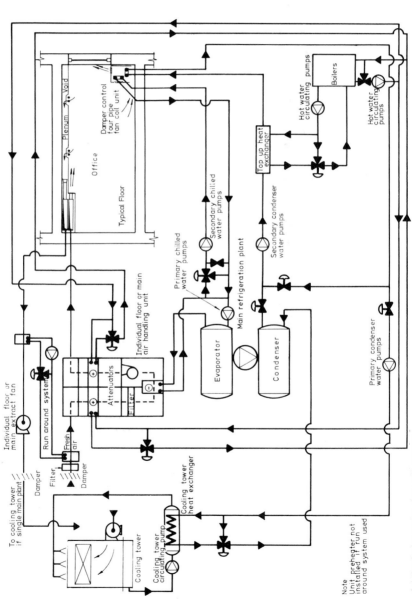

6.4 Basic four pipe fan coil system

illustrates in block schematic form the basis of the system. The following advantages are suggested and illustrated by the fact that there has been less difficulty in letting buildings incorporating four pipe systems.

1. Each window module is equipped with one unit which is ideal for partition changes especially thermostats. As each unit is fitted with a damper actuator pneumatic air lines can easily by changed to suit any partition arrangement. Depending on layout one thermostat serves a maximum of four units. Fig 6.5 illustrates the mode of operation for the fan coil unit.

2. Units can be designed within specific acoustic criteria for commercial developments. Fig 6.6 illustrates acoustic performance at two speeds – specific criteria are more likely to be achieved as there is no problem of balancing high pressure air supplies and associated noise regeneration.

3. Speed control can be applied to each unit to suit the sensible heat loads of each individual office.

4. The existing heating pipework can often be used which only leaves chilled water pipework to be installed for perimeter units together with electrical supplies.

5. Application of heat reclaim techniques is possible, the large heating coil surface allowing use of low temperature water.

6. Units can be switched off when not in use, examples being conference rooms, board rooms etc. If some tenants require to work overtime, the time clock on those floors can be overriden with minimum local and central plant operating.

7. Only cleaning of lint screens and coils is necessary – there are no nozzles to block up as may be the case with induction units.

8. The fresh air is supplied from individual floor air handling units, which in a fire would minimise the spread of smoke since each floor comprises a separate fire compartment.

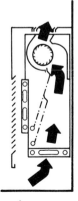 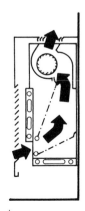 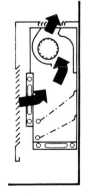

maximum
heating

bypass

maximum
cooling

6.5 Damper sequence of four pipe fan coil unit

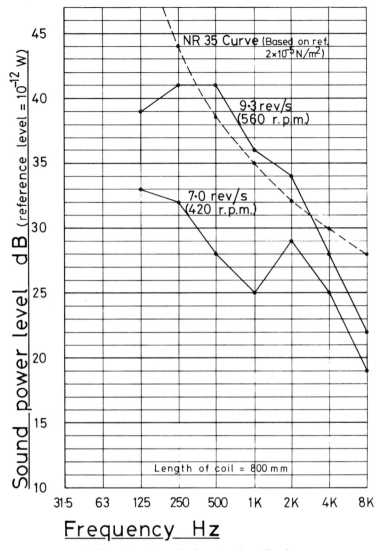

6.6 Sound spectrum for four pipe fan coil unit

9. No recirculation of primary air is necessary and latent cooling can be arranged from the individual air handling units. With a four pipe system full advantage can be taken of primary air credit.
10. Subject to selection of air handling plant, fan and coils, the total primary air supply to each floor can be adjusted to suit occupancy – thus saving on electrical costs and fuel.
11. Each floor can be individually balanced on the air side thus saving time during the commissioning stage and individual floors let, saving on operating costs during the letting period.

12. Preheating with unit fans in operation can be used thus allowing an optimiser system to be used for maximum fuel economy. For a recently completed building with large glass areas the preheat time is only 1¼ hours at −1.0°C external air temperature.

13. Extract air through the lighting fittings can be easily arranged to lower the room sensible load and not pass unwanted heat back to the main plant.

14. Since both cooling and heating pipework is provided at each floor level supplementary air conditioning can easily be provided in for example board rooms, assuming some allowance has been made in the distribution system.

15. Only primary air ductwork is required (sized for fresh air per occupant and latent cooling). This can be low velocity for energy conservation and accommodated in restricted false ceiling spaces.

16. Free cooling of the chilled water in winter can be fully used with no cancelling of excess reheat.

17. With the introduction of sprinklers for large commercial developments (especially requirements within the GLC area under section 20 of the London Building Acts) co-ordination of services will certainly be easier than say variable volume or all air systems.

18. Where only very limited plant space can be constructed (for chillers, cooling towers and additional water storage tanks) the system will occupy the minimum of existing space thus gaining some rental advantage.

19. Experience has shown that the four pipe *damper* controlled unit responds very quickly to loads imposed on light weight highly glazed buildings.

20. The pipework can be designed on a simple reverse return basis with no complicated two or three port control valves.

21. Using kW hour meters costs per floor can be established for service charges.

Variable air volume and four pipe induction systems

The four pipe fan coil system is not the only air-conditioning system useable for refurbished buildings. Both variable air volume and four pipe damper controlled induction systems have been employed.

The variable air volume system is very acceptable if adequate space is available above the false ceiling, say, at least 300 mm clear. It is very important to ensure that some form of perimeter heating or re-heating on the air side is provided otherwise very severe problems can result. A recent completed project for a banking hall has proved very successful but heating was used on the air side. For this project a large void above the false ceilng was available.

For another refurbished building only 205 mm was available within the false ceiling space which was sufficient to accommodate pipe-work and horizontal four pipe fan coil units for internal zone requirements. The perimeter walls of the building (between floors and vertical support columns) were demolished and replaced with a light weight aluminium cladding. Within the cladding vertical false columns were constructed to accommodate high velocity ducting and pipework for perimeter four pipe damper controlled induction units. Again the building was let within a reasonable period of time taking into account poor market conditions at the time.

Client instructions and information

When air-conditioning systems are considered it is perhaps unfortunate that the client does not always give clear instructions to the professional design team. In some cases a professional valuer experienced in rental matters will advise the design team. The following is a short check list of information required.

- What capital allowances are possible?
- Are other grants available from the Government?
- Will the building be occupied by one tenant or a number of tenants?
- Is overtime likely to be worked?
- Is each tenant to be charged separately for air conditioning?
- If the letting market is difficult will the letting be made on a floor by floor basis?
- Is a short construction period important?
- Are record drawings available for the building including structural and services?
- Will false ceilings be allowed?
- What height floor to ceiling does the architect or client require? Will this leave sufficient space for services if a false ceiling is required?
- Are operating costs important?
- What space will be available for plant such as cooling towers, refrigeration plant, pumps and additional cold water tanks? For example, will space be available for some plant in an existing car park? Will the structure allow cooling towers and tanks at roof level?
- Is it possible that the local authority will require sprinklers?
- Will a tenant require maximum flexibility for partitions or will open plan be the basis of design?
- What design specification is required for the services? Is first cost the most important ie speculative development?
- What rental levels are likely?
- Will the design team be required to provide the client with estimated operation costs?

ENERGY CONSERVATION MEASURES

For existing buildings the following could be considered:

1. Fix aluminium foil faced thermal insulation to the inside of external walls (at least 25 mm thick) behind perimeter air conditioning units.
2. Solar reflecting film can be applied to single glazed windows.
3. Consider double glazing if the client is interested in long term financial returns. R. M. E. Diamant (1) appears to suggest that in certain circumstances an economic case can be made for double glazing. For summer conditions blinds between the panes will be far superior for solar rejection. For buildings with single glazing and blinds there is no reason why a supplementary sliding window system cannot be installed inside the existing glazing. Calculations carried out based on current prices with a discounted cash-flow technique suggest that double glazing of the type described should be economic within a 30 year period, the discount rate used being 10%.

The real question is what is the likely effect of energy costs. The Government Paper *Energy Conservation* (2) presented to Parliament July 1976 suggests that 'further increases in the prices of at least some of the fuels will be needed in order to bring them up to fully economic levels'.

A further Government publication *Gas and Electricity Prices* (3) presented to Parliament May 1977 again makes the point 'The Government considers that gas and electricity prices should be at economic levels which reflect the cost of supply, encourage the best use of national energy resources and avoid subsidies from public expenditure. There have also been other official pronouncements that the Department of Energy expect prices to *double* in real terms by the end of the century.

This information and Government statements give added weight to the suggestion that an economic case can be made for double glazing if a long term view is taken. It does seem rather unfortunate that for commercial buildings in the UK capital allowances are not permitted for expenditure on double glazing or structural thermal insulation, whereas allowances are permitted for capital expenditure on the larger boiler or refrigeration plant to consume an increased quantity of fuel for a poorly insulated building.

Future building regulations may require a high level of thermal insulation for commercial buildings. A start has been made for domestic buildings and the professional design team may well make this point when reporting to their client.

4. Provide kW hour and maximum demand meters for large items of equipment such as cooling towers and refrigeration plant. Used together these two meters can show periods of high energy use and peaks. During spring and autumn the load on refrigeration plant can be

reduced for very short periods to reduce the maximum demand. Used properly, meters such as these can help with efficient energy management.

5. Two stage pumping for heating and cooling hydraulic systems. Two pumps can be selected so that when operating together give maximum duty. With one pump operating the flow rate will be reduced to, say 50 to 70% of maximum. Examination of thermal output/flow rate characteristics for coils will show that the relationship is far from being linear. The reduced flow rate will probably serve the building for 70 to 80% of its operating period. Why run one pump at maximum flow rate for the full operating period consuming unnecessary energy.

6. Low temperature condenser water from refrigeration plant can be used for heating at both air handling and perimeter terminal units. It is important that the condenser water is not used directly but heat exchange arranged by indirect methods.

7. Consider power factor correction since air-conditioning plant such as refrigeration machines operate below their maximum duty for most of the operating period.

8. Relate the boilers to the load. Multiple boiler installation can show distinct advantages. Work by Spivey and Jarvis (4) deals with this topic in some detail.

9. Consider very carefully the relationship between capital cost and pressure differential for all items of equipment. Plant with high pressure differentials are very often associated with lower initial cost but throughout the operating period of the plant very high operating cost can result. A comparison between orifice plate and venturi method of fluid measurement for example is given in reference (5).

10. Run-around coils or thermal wheels between extract and supply air systems can also be considered.

The above list is limited to a few of the more important heat conservation measures and is not intended to be complete.

A sound financial basis should be applied to any energy conservation study and techniques developed by Carsberg and Hope (6) for applying inflation to discounted cash flow calculations are very useful in making an appraisal. Capital, fuel and labour costs are all involved and it may well be important to consider each element of cost since their individual relationship to general levels of inflation will be different.

A BRIEF CASE STUDY OF TWO REFURBISHED BUILDINGS

Both buildings were of lightweight structures with high glazing levels. One of the buildings underwent a complete change in appearance with the exterior shell (apart from the structural components) demolished and replaced by a new lightweight metal structure.

Table 6.1 **Schedule of characteristics and data for buildings A & B—considered to be lightweight construction with large glazing areas**

Item	Building A	Building B
Air conditioning system	Four pipe damper controlled fan coil units – no internal zone: Individual floor air handling units	Four pipe damper controlled induction units – perimeter: Four pipe water controlled fan coil units – internal zone Central air handling plant
No of office floors	15	7
Glazing: Blinds or curtains by tenant	Single – with 3M-P33 Scotchtint	Double – float and antisun bronze
Approx: Net lettable floor area Total Offices	6 274 m^2 5 968 m^2	2 930 m^2
Space in false ceiling for services	120 mm	205 mm
External conditions: winter ,summer	−1.0°C dry bulb 100% saturation 30°C dry bulb 20°C wet bulb	−1.0°C dry bulb 100% saturation 30°C dry bulb 20°C wet bulb
Internal conditions	20°C to 22°C dry bulb: 50% to 38% saturation. Swing to 24°C at peak periods	20°C to 22°C dry bulb: 50% to 38% saturation. Swing to 24°C at peak periods
Allowance for office machines	10 watts/m^2	10 watts/m^2
NR level in lettable office area	35	35
General lighting level	300 lux	400 lux
Density of occupation for initial design	1 person 5.5 m^2	1 person 5.5 m^2
Fresh air per person for initial design	0.0118 m^3/s	0.0118 m^3/s
Height floor to underside false ceiling – lower under bulkheads	2.515 m	2.515 m
Low brightness air handling lighting fitting serving office area	No. off 1 532 Total load 150.6 kW	No. off 833 Total load 80.885 kW
Number of terminal units	828 perimeter fan coil units @ 35 Watts each = 29.0 kW total	238 perimeter induction units 86 internal fan coil units = 8.712 kW total
Humidification	None	Spray cooler coil Motor load 4 kW
Boiler capacity	3 off equal capacity 1 758 kW total	10 off equal capacity 700 kW total
Refrigeration plant	3 – reciprocating total capacity 1 153 kW. Total motor load 400 kW	2 – reciprocating total capacity 647 kW. Total motor load 197 kW

Item	Building A	Building B
Offices only:		
Fresh air	14.06 m³/s	7.9 m³/s
Extract air	14.11 m³/s	8.0 m³/s
Cooling tower forced draught	Power 37 kW	Power 11 kW
Condenser water pumps	2 off both duty	2 off both duty
	Motors 2 × 22 kW	Motors 2 × 4 kW
Secondary chilled water pumps		
Perimeter	2 off both duty	2 off duty & standby
	Motors 2 × 22 kW	Motors 2× 7.5 kW
Internal zone	None	2 off duty & standby
		Motors 2 × 3.0 kW
Primary chilled water pumps	2 off both duty	2 off both duty
	Motors 2× 11 kW	Motors 2× 7.5 kW
Primary heating pumps	3 off two duty	2 off duty & standby
	Motors 2 × 1.1 kW	Motors 2× 3 kW
Secondary heating pumps		
Perimeter	2 off both duty	2 off duty & standby
	Motors 2× 5.5 kW	Motors 2 × 1.5 kW
Internal zone	None	2 off duty & standby
		Motors 2× 1.1 kW
Air handling	2 off both duty	
	Motors 2 × 7.5 kW	Part of primary
Air compressors	2 off duty & standby	2 off duty & standby
	Motors 2 × 7.5 kW	Motors 2 × 3.7 kW
Primary air plant	29 individual units	1 off
	× 0.55 kW = 15.4	Motor 1 × 37 kW
	1 × 1.5 kW = 1.5	High pressure plant
	Total 16.9	
	Low pressure plant	
Extract air plant	2 off both duty	1 off duty.
	Motors 1 × 18.6 kW	Motors 1× 15 kW
	1 × 22.4 kW	
General vent plant and miscellaneous services supporting the various systems	Total 35 kW Includes power for oil burners	Total 12 kW Gas fired

The problems for both buildings were similar, namely, very restricted space in the false ceilings, valuable basement storage area required for plant, and space required for cooling towers and tanks at roof level.

It was impossible to consider an all air system such as variable air volume and in both cases air/water systems were employed. Table 6.1 is a schedule of the basic characteristics and data for both buildings. Building B was very much easier to deal with since both a new core and exterior were constructed and space for services could be redesigned. For both buildings no heat recovery plant was installed since financial pressures were such that only a basic system was possible.

Options are still open for heat recovery plant to be installed since both designs incorporate four pipe terminal units which could use, for example, low temperature water from the refrigeration plant condensers.

Where financial pressures are necessary it is very important to leave options open by designing plant to be easily modified when future finance is available or basic fuel/energy costs rise above the general level of prices and inflation.

Such a change has already occured as comparison of changes in the index of retail prices and cost of electricity from a maximum demand LEB tariff for a large air-conditioned building will show.

Date	Cost of electricity pence/unit	Index of retail prices
Sept 1977	2.3	185.7
Sept 1974	1.2	111.0
Change	1.1	74.7
% Increase nearest whole number	92	67

These increases in electricity costs above general price level increases and recent work by D. J. Fisk (7) suggest that designs should allow easy modification incorporating the idea of option value. A factor which was very much taken into account in dealing with Building A.

The building consisted of ground to 15th floor, the latter being a restaurant and kitchen area. The ground to 14th floor consisted of offices with a central core and occupancy was required on the 8th to 14th floors before the building was complete. Hardly any space was available in the existing core with the false ceiling height being only 120 mm. Space for pumps and refrigeration plant was limited at basement level and only the very minimum was allowed since car parking spaces were at a premium.

Plant space was limited at roof level due to weight restrictions and planning problems associated with plant room height.

The problems were solved by using the existing perimeter heating system and installing four pipe damper controlled fan coil perimeter units. Chilled water was provided from the central refrigeration plant at basement level via vertical distribution with ring mains in the false ceiling serving the individual units. Treated air was provided from two small air handling units at each floor level with distribution ductwork around a core bulkhead with grilles delivering the air into the respective offices. Holes for services were only required for the chilled water and heating pipework which kept within the concept of minimum alteration to the structure. The extract air from the offices was via the lighting fittings into a ceiling plenum with a high velocity common extract system within the core split between 8th to 14th and ground to 7th floors. This arrangement allowed for occupation of the upper floors before final completion. Apart from the extract system, central heating and chilled water distribution, the air side on each floor was self-contained which made commissioning very simple and easy to carry out. The air quantity to each floor could be modified to suit the individual floor occupancy level by simple changes of fan pulleys within the air handling units. The various chilled water and heating systems were each fitted with two pumps – one pump designed for operating most of the year at around 60% of the maximum duty and the other pump only being required at peak conditions. This assisted with the initial occupancy of the building.

The existing oil fired boilers were retained. A very large margin had been applied to the previous heating load which was more than adequate to meet the heating loads of the air conditioned building. It is not surprising that the present installation was adequate as the four pipe design incorporates low water temperature varying between 48°C and 20°C.

Previous experience had shown centrifugal refrigeration plant to be unsatisfactory and the engineering policy to install reciprocating machines was continued. This was further influenced by the car parking constraint which forced the chiller room to be long and narrow. The refrigeration plant had to fit in a one metre width and still allow space to provide heat recovery plant if this proved economic in the future. The pumping system used very little floor space – approximately 33 m².

The cooling tower was situated at roof level and to assist with design loads the conditioned extract air from the 7th to 14th floors was diverted into the cooling tower chamber.

The requirements of the Fire Brigade officers and the District Surveyor were not onerous since no recirculated air was employed and fire dampers were limited to the extract only, a very important fact when a partial occupation consent is required from the GLC.

CONCLUSION

Appendixes 6.1 and 6.2 give detailed information about fresh air rates and air conditioning costs. Fresh air rates can have a considerable influence on operating costs and must be carefully considered. Keeping an air-conditioning plant simple is also important since minimum staff can be employed.

During the latter part of 1977 buildings A and B have been partially occupied but actual operation costs were not available at the time of writing.

What of the future? It seems likely that air conditioning will survive since our economy should become progressively stronger. The negotiating position of the larger unions may well exert an influence since they will require better working conditions for their members.

Air conditioning is expensive to install and operate. Research and energy management must be strengthened to allow costs to be kept within reasonable levels.

REFERENCES

1 R. M. E. Diamant, *Insulation Deskbook,* Heating and Ventilating Publications Ltd, 1977.

2 Department of Energy, *Energy conservation,* The Government's Reply to the First Report from the Select Committee on Science and Technology, Session 1974–5 HC 487, presented to Parliament, July, 1976.

3 Department of Energy, *Gas and Electricity Prices,* The Government's Reply to the Fourth Report from the Select Committee on Nationalised Industries, Session 1975–6 HC. 353, presented to Parliament, May, 1977.

4 H. B. Spivey and D. J. Jarvis, *Modular Boiler System for Commercial Heating and Hot Water Supply,* The Institution of Gas Engineers, 1973.

5 P. H. Day, Capital Investment Appraisal for Mechanical and Electrical Services in Commercial Buildings, *The Heating and Ventilating Engineer,* February, 1976.

6 B. Carsberg and A. Hope, *Business Investment Decisions under Inflation,* The Institute of Chartered Accountants in England and Wales, 1976.

7 D. J. Fisk, *Energy Conservation: Energy Costs and Option Value,* Building Research Establishment, Department of the Environment, CP 57/76, July, 1976.

8 Notebook: Recommended outdoor air supply rates for air-conditioned spaces Table A9.24 (revised November 1971) *Journal IHVE,* March 1972, volume 39.

9 Ventilation Requirements Fig 4 for ceiling height of 2.7 metres, *Building Research Establishment Digest 206,* October, 1977.

10 Medical Officer of Health, Port and City of London, various private correspondence for projects in the City of London dealing with fresh air ventilation rates.

11 GLC London Building Acts (Amendment) Act 1939, section 20, fresh air ventilation rates as part of actual consent document.

12 G. W. Brundrett, *Ventilation requirements for smokers,* Electricity Council Research Centre, ECRC/M870, December, 1975.

ACKNOWLEDGEMENTS

To the Directors of The City of London Real Property Company Ltd for allowing the paper to be presented and the use of Company information. Any opinions expressed are my own and not necessarily those of the Company.

To Weathermaker Equipment Ltd and Atholl Engineering Ltd for details of the four pipe damper controlled fan coil units.

APPENDIX 6.1 FRESH AIR RATES

Table 6.2 Basic data for fresh air rates – l/s per person

Authority	Offices – open plan	Offices – private	Offices – exec.
(8)	Minimum 5	Minimum 8	Minimum 18
IHVE	Some smoking	Heavy smoking	Very heavy smoking
CIBS	Recommended 8	Recommended 12	Recommended 25
	Some smoking	Heavy smoking	Very heavy smoking
(9)			
Building			
Research			
Establishment	4.3	7.08	
Digest 206	Some smoking	Smoking	
(10)			
City of	Minimum 9.4 ⎱ No reference to smoking or		
London	Recommended 11.8 ⎰ type of office		
(11)			
GLC London	21.236m³/hr (5.9 l/s) per person ⎫	Wording taken from	
Building	or per 4.572m³/m² of floor area ⎬	actual consent. No	
Acts,	whichever requires the ⎭	reference to smoking	
section 20	greatest ventilation	or type of office	

For some air/water systems such as two pipe non-changeover induction the amount of primary air is associated with rates to suit unit sensible cooling performance, room latent loads, adequate heating and acoustic levels.

These requirements will probably not be met if only fresh air of a low quantity is used. A greater volume of air with some recirculation may then be necessary associated with larger ducts, area of insulation and smoke detection in case of fire.

For refurbished buildings space for services is at a premium and primary air ductwork must be kept to an absolute minimum.

Examine the position for a four pipe damper controlled fan coil unit system. Assume a minimum office volume of 11.33 m³ with one occupant and a summer design condition of 22°C dry bulb and 50% saturation. Allowing for infiltration of external air a room latent load of approximately 59 watts would be a typical value. A room supply air moisture content of approximately 0.006673 kg/kg of dry air would be necessary for an air flow of 11.8 l/s.

With this system only latent load requirements need be considered initially but a judgement must be made if 11.8 l/s will be used as a fresh air rate/person to avoid any recirculation and its associated complications.

Examine the chart prepared for fresh air rates suggested by various authorities and it will be observed that the calculated figure of 11.8 l/s meets the City of London requirements.

An examination of the psychrometric chart plot will show that the moisture content of 0.006673 kg/kg of dry air is low and it is doubtful if this should be reduced any further since at that value the cooling coil will probably be 10 rows and a chilled water supply temperature of 5°C will be necessary.

If the City of London fresh air rates are considered unacceptable then recirculation will be necessary if lower rates suggested by some authorities are used.

Consider a floor to ceiling height of 2.515 m for a refurbished building then the minimum floor area per person will be 4.5 m². This density of occupation is possible where junior clerical staff are concerned and high rental levels exist.

A very different situation can exist for a director's floor. An actual value for a recent project was approximately 1 person/20 m² allowing for some visitors at a meeting in the conference/board room.

As the air to this floor was based on the general specified value of 1 person per 5.5 m², and 11.8 l/s of fresh air per person then the total air quantity could be reduced assuming 25.0 l/s of fresh air per person to allow for possible increased smoking activity. Allowance must be made in the figures for space occupied by secretaries when considering the reduction in the fresh air rate to the floor.

So far two extreme conditions have been chosen but for general development the designer does not know what individual floors will be used for and the value of 5.5 m² per person does not seem an unreasonable figure at present.

The main consideration is to have the ability to adjust the plant to suit occupancy levels after the tenant has moved into the building and some level of stabilisation has been reached.

G. W. Brundrett (12) suggests that for large open plan offices only half the occupants will smoke. This poses a question – will the publicity to stop smoking reduce the suggested 50% level?

Clearly the standard specification put forward of 5.5 m²/person and 11.8 l/s per person is not claimed to be a complete answer but an attempt to suggest a solution hoping that this will promote discussion on a very important aspect of air-conditioning design and energy conservation work especially when refurbished buildings are considered with minimum space for services.

APPENDIX 6.2 AIR-CONDITIONING COSTS (CITY OF LONDON) AIR/WATER SYSTEMS

It is very important to establish the actual costs for air conditioning. These are scheduled in four main parts:

- operating
- replacement of main plant and services
- rental value
- local authority rates

1. Operating

The operating costs for air conditioning can usually be divided into the following main elements. Figures are for September 1977. The electrical, fuel and water figures are purchase prices.

1.1 Electrical

For motors and thermal heaters. Typical cost 0.64p/MJ (2.3 p/unit).

1.2 Fuel

For boiler plant or other direct fired equipment. Typical cost: Gas 0.17 p/MJ (17.9p per therm) Oil 0.200 p/MJ (34.0p per gallon).

1.3 Water

Cooling towers and general make up for other systems. Water rate is 5.3% of the net annual value less 50% rebate.

1.4 Contract maintenance for specialist plant

Water treatment.

Controls.

Refrigeration plant.

Cleaning and replacement of nozzles, lint screens for terminal units such as induction and fan coils.

Burners.

Electrical starter panels.

Specific quotations can be obtained for this work.

1.5 Major expendable items

Filters for primary air plant.

Chemicals for water treatment, ie cooling towers, heating and cooling systems.

Chemicals for cleaning heat exchangers, coils, refrigeration machines and the like.

1.6 Insurance

This will vary with the cover required; generally rotating plant will be included and pressure vessels.

1.7 Miscellaneous materials (not complete list, indicative only)

Belts.

Seals.

Fuses.

Cleaning materials.

Lubricating oils.

Tools.

Refrigerant.

Air compressor filters.
Spares for refrigeration and boiler plant.
Bearings.
Spares for control system.
Rags, and general cleaning materials.
Stationery.
Specific quotations can be obtained for these items.

1.8 Labour for general maintenance
Basic salary.
Overtime.
Pension and insurance.
General overheads, including head office and general supervision (also allowance for space occupied by staff which could otherwise be let).
Holiday relief.
Sickness relief.
A typical total rate per person at 1977 costs would be £10,000 per annum.

2. Replacement of main plant and services
There are numerous methods which can be applied but a system which gives a constant charge but reflects inflation is described:

Example
- Consider initial capital cost for plant and services at £80 per m² related to air-conditioned floor area.
- Assume building of 6000 m² net lettable air-conditioned area.
- Assume 30 years average life for plant and services. The 'period' chosen will depend on the standard of plant and services installed and operating hours per annum.
- Replacement of plant and services – revalued at current cost for each year.

1st year cost

$$\frac{6000 \text{ m}^2 \times £80/\text{m}^2}{30 \text{ years}} = £16,000 \text{ per annum or } £2.66/\text{m}^2$$

2nd year cost

$$\frac{6000 \text{ m}^2 \times £100/\text{m}^{2*}}{30 \text{ years}} = £20,000 \text{ per annum or } £3.33/\text{m}^2$$

*To replace the plant and services after one year taking into account inflation etc for that year. This involves a revaluation by a professional engineer to establish the additional charge over the original first year cost.

3. Rental value
This will vary with location and type of building. For the City of London air

conditioning may increase the rental value by £20 per m² related to net lettable air conditioned floor area.

4. Local authority rates

An assessment will be made by the Inland Revenue. From the gross value the rateable value is calculated by the formula

(Gross value £ minus £34) × $^{10}/_{12}$

The local authority then applies a percentage rate which for the City of London is 78.33. The above is standard formula for any rateable value of any gross value above £430. Deduct £34 from gross value, add '0' to balance and divide by 12.

The additional cost for air conditioning is calculated approximately as £6.5/m².

Summary of cost (air conditioning)	£/m²
1. Operating – typical value calculated using charges (1.1 to 1.8) inclusive	11.5
2. Replacement cost of main plant and services	2.66
3. Rental	20.00
4. Local authority rates	6.5
Total for first year:	40.66

General notes

The cost for the operating of lighting should not be part of the air conditioning but a charge under the heading of 'office lighting'. This suggestion is based on the fact that lighting is required for normal use and the cost of the additional thermal load is taken into account when calculating refrigeration loads for the summer and the subsequent electricity consumed by the chillers.

A similar calculation should be made for central heating for the same building. The total cost should then be deducted from the air-conditioning charge to arrive at the true 'extra' cost for air conditioning. The cost will depend on the type of heating system selected but an approximate figure of £36/m² for the true 'extra' cost is suggested.

It may be considered than an air-conditioned building will require less cleaning and general decoration. On balance this may be doubtful since in practice an overall higher standard of housekeeping will be demanded.

A judgement has to be made regarding increased efficiency of staff working in air-conditioned buildings.

The terms rateable value and net annual value are specific to the fiscal system in the UK.

To assist with estimating the annual fuel consumption for office buildings the following data is offered. This data was prepared by A. E. Rose CEng, MIMunE, Chief Engineer, The City of London Real Property Co. Ltd, to whom acknowledgement is gratefully given.

Table 6.3

Building	Air conditioning	Glazing %	Lighting lux	m^2 floor area	Litres per annum fuel oil	Fuel oil ann. consumption litres/m^2 near. whole no.
A	—	35	400	2 880	95 830	33
B	—	40	400	3 112	106 604	34
C	—	30	400	3 855	101 876	26
D	—	30	350	4 877	119 332	24
E	—	35	400	6 131	198 024	32
F	—	35	400	7 711	271 669	35
G	—	40	400	7 766	266 032	34
H	—	30	350	8 082	217 526	27
J	—	60	400	9 708	294 535	30
K	Yes	40	800	11 148	523 699	47
L	Yes	60	500	14 864	690 992	46
M	Yes	80	500	14 957	841 692	56

7 Air-conditioned buildings in the Middle East

N. NOSSAN
G. PANCALDI

INTRODUCTION

To design and execute an air-conditioning installation in a Middle East country is a desire common to everyone working in building services, be they consulting engineers or contracting companies. It is a topical subject which requires deep examination.

The Middle East market for air conditioning is most promising. Air conditioning is a basic necessity of life in the Middle East countries which can afford to pay for it, in marked contrast to countries such as those in Africa and the Far East which have the same needs but whose economy does not enable them to afford the cure.

Commercially the business is desirable and it is being offered to countries which require it: our main purpose is therefore to carry it out as well as possible to satisfy the customer with the right product and to obtain a fair profit as compensation. To achieve this result it is essential to take into account all aspects of the technical solutions to the problem: the installation must be designed not only to achieve specific criteria of temperature and humidity but with particular awareness of its future operation and maintenance, of difficulties in assembly and erection and the availability of spare parts in the country concerned.

In the context of the Middle East it is important to realise that the best technical solution is useless if it cannot be properly constructed on site or if it will never give its best performance because of the absence of spare parts or the availability of skilled maintenance staff. There are therefore four fundamental aspects to the problem: 1 The technical design of the plant; 2 Site organisation; 3 The execution of the installation; and 4 Maintenance.

TECHNICAL DESIGN OF THE PLANT

The design of the plant is usually carried out by the consulting engineer who selects the type of plant most suitable to the particular kind of building being air conditioned, both from considerations of structure (area of glazing, masonry type, plant room spaces available, etc) and the use to which the building will be put (hospital, office building, etc). The consulting engineer, therefore, plays a vital role in the contractor's business since at the very beginning he decides the real direction of the whole work and its final

134

technical results related to the conditions in the particular country.

His technical solutions, his calculations, his specifications, his bill of quantities and his tender drawings are the starting point for the installation which is quoted for and executed by the mechanical services contractor. It is therefore at this stage that the thinking must be directed along the right lines if a first-rate solution is to be achieved which complies with the conditions and characteristics of the customer country.

As contractors we have been called on to produce all kinds of plant with considerable differences in technological content even for roughly similar buildings. The plants are sometimes too sophisticated, sometimes too elementary. We have over the years developed our own point of view on the kind of solution to choose when we are entrusted with the engineering and contracting of a job. We would not of course suggest that this is a final opinion. The following paragraphs outline some aspects of our view.

Standards

Middle East countries are mainly inclined towards the use of British Standard specifications except for some cases where the American standards and the European ones like the German DIN or the Italian UNI are accepted. We believe that a particular national standard should not be binding for the contractor to all materials without exception.

The selection of equipment based on the availability of spare parts in the customer country is undoubtedly good policy and important, but fixed standards for materials like pipes, valves, ducts, etc carry no real advantage and can be expensive if goods to that particular standard are not directly available in the customer country. The prescription of a type and quality of material, thicknesses, types of joint, etc should be sufficient to allow their purchase at the best conditions in the country of origin of the sub-contractor. In this case it is not the problem of finding spare parts, the main problem is the possible delay to the execution of the works. Different pipes, fittings, etc should be allowed in the one installation provided they comply with the specifications concerning quality when they are available on site or in the sub-contractor's country.

Type of installation

The type of air conditioning installation to be designed depends on the technical requirements of the building, on the factors mentioned above such as maintenance and operation difficulties, water availability, ease of availability of spares.

Shortage of water quickly suggests the type of cooling to adopt for the condenser. We have assessed that 95% of the condensers in the small and medium capacity range of air-conditioning installations are air cooled. With higher capacity installations the percentage that are air cooled goes down to 50%, in particular because centrifugal air-cooled condenser units are rarely used.

7.1 King Feisal Teaching Hospital Riyadh, – main entrance

7.2 Part of the central plant at the King Feisal Teaching Hospital

7.3 New Jeddah International Airport

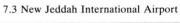

Where there are no skilled staff available to carry out maintenance, the installation has to be as simple as possible. The degree of automatic control must be reduced and not compensated; about 90% of the automatic control systems would be of the electrical type. Large central plants should be avoided unless maintenance is available. It is felt that many simple plants are better than one dual duct system or variable volume system in these circumstances. Induction systems are practically never used.

On the other hand Middle East countries have their personal conception of an air-conditioning installation which is quite different from that in more developed countries. Because of their climate and conditions of life they need a plant providing as much cooling as possible and always running, without any maintenance or repair problems.

The medical conception that too high a difference between outside and inside temperature is detrimental to health is ignored in Middle East countries and often refuted. An inside temperature of 25°C against an outside temperature of 45°C is the best for them. The control of relative humidity is not important. In summer it is sufficient to reduce relative humidity to as low as possible while in winter the problem does not exist.

The use of air velocities which are too high producing noise and draughts does not matter either: it only has to produce cooling. If there are large variations in internal temperature rather than a constant temperature or one that is slightly sinusoidal, it does not matter if it produces cooling.

These are the thoughts of the final customer for whom the consulting engineer is preparing the design, and the consulting engineer must meet the customer's requirements by means of a technology which normally rejects such poor installations as barbarically simplified. An engineer can find such a change of attitude very difficult to achieve, for he will not like putting forward a solution which is technically poor. It is however possible in the Middle East to produce installations with an acceptable technological standard and a satisfactory final technical result which receive the full approval of the engineer as well as the customer. This can be brought about by rationally combining all the above factors. We would give as one example of such a solution the King Feisal Teaching Hospital, Riyadh (7.1 & 7.2), the air conditioning of which was carried out by our company. This is no doubt considered as one of the most specialised hospitals in the world. The air-conditioning installation is a high velocity system using mono duct boxes with reheating coil, and it is fully automatic as is the central plant (steam boilers, absorption units plus cooling towers). An interesting solution has also been adopted for the new international airport in Jeddah, Saudi Arabia (7.3). This is a huge project in which technologically advanced solutions are being used, for example, there is full centralisation of all the cooling and heating systems. These prestige installations are important ones but this degree of technology is very rare. The most common installations in the Middle East which represent the most convenient and suitable installations to them are the ones mentioned below for each building type.

Hospitals

Air systems for ward blocks – good filtration – possible use of high velocity systems to compensate for the distances involved and to reduce the space occupied by the air distribution system – pressure reduction boxes with the reheat coils. Multizone systems for operating theatres and auxiliary spaces – absolute filtration. Use of central plant for producing chilled and hot water – steam boilers and centrifugal or absorption refrigeration plant – possible use of air cooled condenser for centrifugal chillers. The use of minimum automatic control in the wards and good automatic control in the operating theatres.

Offices

Fan coil system, without or with primary air – chillers with air cooled condensers even for large buildings. Air handling units with ducting – hardly any division into zones – roof chillers with air-cooled condensers for smaller buildings. Generally no provision for heating although facilities for electric resistance heating are sometimes provided.

Conference rooms, theatres, etc

Low velocity air-conditioning installations with split system and air-cooled condensing unit on the roof – air handling units with ducting inside. Generally no heating although sometimes electric resistance heating.

Hotels

Fan coils without primary air in the rooms – air handling installations in the corridors and circulation spaces – roof mounted refrigeration plant with air-cooled condensers. Again there is generally no heating.

Obviously, from the point of view of technical excellence, these installations are not satisfactory but they are easy to maintain and operate and are most suitable for the present stage of development of the country. In the future, as development proceeds and the expertise available within the population of the country increases, it will be possible to introduce more complex equipment and use it to obtain better results.

SITE ORGANISATION

When the design and drawings are complete and the difficulties of materials approval are overcome, it is essential to organise the site in advance of the delivery of the materials. Well-organised technical staff and good organisation of the site works ensures a good end product, properly executed within the overall plan to the mutual benefit of the client and the contractor.

Language is vitally important. The technical and managerial staff working on site must speak the official language of the contract, in addition to their native tongue. In the majority of cases this language is English. Knowledge of the contract language must be complusory to avoid the misunderstandings that would arise if technical and commercial discussions had to be translated between the people concerned.

The second essential requirement is to have staff who are properly qualified and have the necessary skills appropriate to the job in hand. The problems to be solved and agreed upon are immediate and decisions have to be taken independently by people aware of and able to evaluate the economic and technical implications.

Consequent upon these considerations our experience suggests two main types of site organisation, although the difference between the two is only related to the way design decisions are taken and documented, whether this is done on site with working drawings produced on site or whether these functions are undertaken at head office.

The former method is rare and more difficult because of the necessity of sending technicians and draughtsmen to site. On the other hand it provides great technical advantage. If the whole of the design/construction team – the architect, the consulting engineer, the main contractor and the sub-contractors – have design offices on site, all the problems of choice of materials, approvals, drawings, variations, etc can be solved immediately and misunderstandings avoided. Such a solution is adopted only for big projects because of the high cost of housing and infrastructure. Further-more, all the organisations concerned must move some of their technical staff from their head office to the site location and this can damage the internal structure of their company, reducing the availability of qualified staff at head office. Where the design work is carried out at head office it is still necessary on site to provide for materials and working drawings besides the installation of the plant.

We believe the following staff provide the minimum organisation essential to a proper execution of the contract.

Site manager – who is responsible for the whole of the contract.

Site supervisors – the number of site supervisors vary of course depending on the size of the contract. They supervise and are responsible for the erection and construction of the equipment controlling the working terms and the staff.

Storekeeper – a very important man. When working abroad it is important at any one time to know exactly what materials and tools are in or out of stock and whether the amount of stocks are sufficient to allow work to continue. A sudden unforeseen shortage of a fitting or tool can cause delays of weeks as it is usually not available on the local market.

Site draughtsman – their job is to produce as-built drawings during the construction period (not at the end as they usually do when all false ceilings and masonry works are installed).

Site engineers – site engineers control the contract from a technical point of view. They are responsible for variation orders, monthly certificates, etc and attend the site meetings and control the planning. Their quantity depends on the size of the job.

To the above list of people must be added the administrative and local staff. The provision of this full staff is expensive and sub-contractors are usually unwilling to spend the amount of money involved, mainly because in producing his tender each has tried to reduce this kind of site cost to a minimum in order to be competitive.

The consulting engineer should ensure that as part of the tender document the provision of site staff is a compulsory item included in those PC items which are to be paid for initially. As fixed price contracts are almost compulsory in the Middle East, the organisation costs should be handled separately from the normal clauses which cover the fixed price contract. They should be related to the duration of the site works so that there is a possibility of a re-valuation where handing over is delayed.

It is an open secret that the terms foreseen in the tender document can almost never be complied with in the Middle East, irrespective of the wish of the contractor to do so. The careful planning to meet the contract date is so often compromised at some stage during the construction or even at the end of the work by delays in clearing imports and transporting goods to site, difficulties in local permits, delays in civil engineering work being carried out by a local firm, etc. Of course, these inconveniences should be allowed for when specifications are drawn up, but so often local laws or authorities do not allow it.

In October, 1977 we were very pleased when we were able to hand over as planned seven sites at which we had built domestic accommodation for the Royal Saudi Air Force. These sites are all in Saudi Arabia at Khamis Mushayt, Abba, Dharan, Riyadh Airfield, Riyadh Soc. and Tropo, Afif, Hanakya. We were very fortunate to be able to achieve this handing over on time and we will scarcely be able to achieve similar results again. As it was, we had to pay a price to achieve it, although it was an economic one.

As contractors, we do not believe we should carry the responsibility of meeting deadlines when we are often prevented from doing so by others so that to have any possibility of achieving the deadline requires every kind of effort. This is deeply unjust and should be officially foreseen and rewarded by including appropriate clauses in the documents.

These comments may seem out of place in a section describing good site organisation but they are made because it is important to realise that even the best organisation may be frustrated and negated by the factors described.

EXECUTION OF THE INSTALLATION

In this third phase of the work the type of labour and its correct use are fundamental. The least risky approach is to engage all specialised ex-patriates but this is expensive. Engaging all local labour is cheaper but contains a considerable element of uncertainty. It is our opinion that the right solution falls between these two extremes, with the quantity of expatriate and local labour and their functions dependent on the particular contract and its sophistication and also on the origin of the contractor.

A particular example is an installation which was carried out in the Middle East in co-operation with an English contractor. Here we used non-working foremen who were expatriate and local labour only for erection. The local labour were given a period of training and after that the foremen directed the work. The result was excellent although the quality of finish of the plant left a little to be desired. Some European companies have used only Pakistani or Korean labour but the technical results have been poor and the programmes tended to fall behind target.

We prefer to carry out our installations with at least 80% of expatriate *working* foremen and 20% of Asian workers using also some local help. With this balance we are certain that when installed the plant will work successfully and that the level of finish will be good. Also if no unforeseen circumstances intervene the job will be ready on time. It is not essential that all the foremen speak English but they must be highly skilled workers able to cope with the rather heavy climate and to work on their own initiative. Of course the percentage of expatriate local labour may vary during the course of a contract depending on the particular kinds of work there are to do at any one time. It is vital that the completion of the job and commissioning and tests are carried out by expatriate skilled labour. These are very important and delicate phases requiring technical skill which is absent in the local labour. All the equipment, especially the delicate pieces, must be handled in the right way if damage is to be avoided, either at the time or in the future. This is particularly true where incorrect installation or assembly of one component could create damage to the whole or a large part of the installation. We prefer not to use skilled labour from manufacturing companies and equipment suppliers. We prefer our own skilled staff who know and can identify the problems of particular items of the equipment and can also relate them to the design and direction of the whole plant.

The final phase – the handing over of the installation to the owner – may itself entail losses of time and disagreements between the sub-contractor, the main contractor and the consulting engineer. The procedure may vary according to the country but everyone wants to leave the site as soon as possible leaving all the difficulties and problems to others.

The normal English procedure of checking the quantities and the quality of the installation with the consequent issue of a 'snag list' at the provisional acceptance is the best one. With provisional acceptance carried out in this way, the contractor is obliged to execute repairs contained in the 'snag list' within the guarantee year. During this year also the contractor is responsible for equipment and plant defects. The final handing over occurs at the end of the year and at this time the responsibilities of the contractor cease.

This simple proceeding which is by no means always adopted should be clearly required in the specifications. If this is done, misinterpretation at the end of the work is avoided.

MAINTENANCE

This is the most critical phase of supplying and operating an air-conditioning system, both in the way it is usually requested and in the way it is carried out.

During visits to Middle East and African countries we occasionally see air-conditioning plants originally installed by us. It seems that completed installations are rapidly neglected and where installations are inadequately maintained they are destined for quick destruction by natural agents.

'Maintenance' itself is a word which when translated into other European languages has rather differing meanings. If it is translated into Italian with the word 'manutenzione', the operation of equipment is excluded. If it is translated into French with 'manutention', it means moving of the materials inside the site. The corresponding word really is 'entretien' but even this includes other activities as well. Maintenance as a description is not sufficiently specific and it is important to specify exactly what activities are to be undertaken by the contractor or the owner. The specification should also allow for the plant to be operated by the contractor at least for the first year.

There are really two acceptable ways in which maintenance can properly be effected. In the first method the contractor is responsible for guaranteeing the installation only after he has trained staff for maintenance and operation of the plant that are employed by the owner. This is the least expensive of the alternatives to the owner but it is important that the owner's maintenance and operation staff start their training with the contractor at the beginning of the installation period. They should work physically on the site for the whole of the installation period, attending all the completion work, tests and the commissioning. The contractor should have the right to ask for new staff to replace anyone judged unsuitable. The alternative proposal is for the contractor to be responsible for the first year's full maintenance and operation of the plant as well as for its installation. This responsibility for maintenance and operation will include direction and supervision but not doing the job itself, which will be undertaken by staff employed by the owner. The contractor must therefore supply essential staff, such as the foreman of the operations team and the foreman of the maintenance gang as well as the specialists to deal with particular items of equipment, eg chillers. The operation and maintenance teams supplied by the owner will during this first year learn all about the plant and how to carry out the required maintenance operations. Even at the end of this period, this team of technicians will not be able to deal with all problems. It will always be necessary to have foreign specialists in for some items of repair and maintenance work, although the owner's workforce trained in this way will be able to deal satisfactorily with the majority of the work required.

In our view the above alternatives are the only ones if the plant is to be operated in the best possible way. Sometimes in the Middle East a third more expensive solution is adopted whereby maintenance and operation of a plant are carried out by skilled foreign companies. This method may be the best technically but will do nothing to assist the technical development of the country.

CONCLUSION

Air conditioning in the Middle East is a very important subject which could have been dealt with in different ways. Obviously the technical aspects could have been developed to a greater extent than has been done here by detailing the various possible solutions and giving examples of installations with data on costs, etc. Instead, we have tried to highlight experiences and conclusions obtained from working in the Middle East, trying to provide better building services with a good technical and economic performance.

As the final point, we would like to mention the possibility of application of solar energy to the Middle East. Both European and American industry are producing improved solar panels but so far these have not been adopted or used to any great extent in Middle East countries. It might be thought that this is because they have sufficient cheap oil energy to make solar energy unprofitable and superfluous in spite of the climatic conditions which are extremely suitable to making use of solar heat.

The real Middle East problem is the demand for electric power which as the construction programme develops grows year after year faster than the number of power stations to supply it. It is for this reason that there would be considerable advantage to be gained in reducing electricity demand by heating domestic hot water using solar panels. It seems highly likely that such a use would be economic and certainly deserves serious study.

DISCUSSION

G. Mole (Bernard Sunley & Sons Ltd)   There is one thing Mr Pancaldi did not emphasise sufficiently, the problem of corrosion in the Middle East, although he did say that plant should be sturdy. We find in the Middle East that as soon as bare metal is stored it corrodes very rapidly – a problem not understood by many British manufacturers.

My company has a different philosophy than Aster International on the way we run our sites. As Mr Pancaldi said, they like to have people actually running the job from the site. We believe that people on the site have quite enough to worry about with the poor quality of labour, the poor weather conditions etc, so, where possible, we like to do all our ordering from the UK and have all the drawings prepared and approved in this country so that when the site receive them it is a case of installing to those drawings rather than worrying whether things are right or not.

G. Pancaldi (Aster International Spa, Milan)   It is perfectly true there are advantages in using UK products and designing in the UK but there are also advantages in undertaking these functions on site. Admittedly it is easy in an English-speaking country like South Africa where having architects and consultants on site problems will be quickly solved. If you design in the UK and people in Dubai find the drawing does not correspond to the real construction – quite a common problem – there are real difficulties in putting it right. We do not carry a large staff, sometimes only using local contractors. If design work is done at the site problems will be solved quickly. This arrangement produces just one problem – all your staff are away from the head office so cannot be used on other work.

L. J. Wild (George Wimpey & Co Ltd)   I endorse Mr Pancaldi's approach to work in the Middle East which is similar to that of my own company. The corrosion problem referred to by Mr Mole depends on the humidity of the location. A site near Esfahan had to be closed due to lack of forward funding. When we returned 18 months later the structural steelworks were standing and there was no corrosion. This was a high altitude site with low humidity.

8 At the end of the day: an optimist's view of energy conservation

JOHN P. EBERHARD

SHORT HISTORY OF BUILDINGS

In discussing the relationship of architectural design to energy conservation it would be useful first to review a short history of buildings. Especially to look at the various stages of building technology and to examine the relationship between buildings as modifiers of climatic conditions in order to accommodate human activities in as comfortable conditions as possible.

In our earliest forms of human habitation, man often lived in spaces found in nature – such as caves or in holes in the ground covered with light materials. In such solutions the natural thermal protection of the earth's crust provided a levelling mechanism against the temperature excursions outdoors. This principle is now being re-explored under the general notion of earth berming or even underground dwellings.

The development by the Eskimo of the igloo, by the American Indian of the tent, and by the native African of the hut were all the result of utilising easily available raw materials and simple technologies to provide a means of shelter. While not the result of a conscious design effort each of these managed to be quite sophisticated solutions to the climatic conditions in their part of the world. Each of these original cultures departed in this century from these traditional solutions to use more 'modern' materials such as corrugated metal roofs or tar paper walls on flimsy wood structures. Wherever this was done the resulting dwelling was not only much less comfortable than the traditional solution, it was far less handsome in a design sense.

In a few places in the world there evolved solutions that were even more sophisticated in terms of climatic adaptation. The housing complexes of the Pueblo Indians in the Southwestern United States are an elegant response to the sun's pattern of movement during the year – each of the dwellings is exposed to the beneficial heat of the sun during the winter months and protected from the excess of the sun's heat in the summer months. The 'wind scoops' of historical dwelling units in Hyderabad, Pakistan were extremely inventive passive solutions to providing cooler conditions in a very hot climate, however they provide an obvious excitement as a design element as well. In both of these cases the sophistication of the climatic response was the clear generator of architectural designs that were elegant and handsome.

145

One of the most beautiful solutions to human habitation that is clearly related to functional purposes which have been completely merged with climatic responses are the tents of the nomadic peoples of Arab descent. With no identification of individual designer's record, those tents represent a design solution of supreme elegance – combining colour, texture, engineering, fabrication, and utility with pure delight.

EXAMPLES OF ENERGY ADAPTATION

If we move further along on the historical avenue of technological development we find some examples of energy adaptation that are so well designed that they are in and of themselves classic items of design excellence. These include the windmills of Holland, the water wheels of New England and the Scandinavian stove. While these three examples are devices for converting one form of energy to another, notice that they are intended to be seen (not hidden) and to be admired as well as to be effective energy convertors (almost the opposite of those heating devices which Michael Brill has characterised as 'the monster in the basement' of most American homes).

The fireplace

In the fireplace, another device intended in a utilitarian sense to provide a means of converting energy from fuels into heat (and even light), we can find a long history of design development. The basic principles of providing a contained space in which to burn fuel safely, to exhaust the unwanted smoke, and to radiate heat into the surrounding space have not changed dramatically over the centuries. The fireplace seems to have its origins in the 11th or 12th century in response to providing a more controlled and safer fire within buildings and the companion problem of ridding the interior of the smoke and soot that hung in the air of ancient castles and cottages.

By the 18th and 19th century the fireplace was an established artifact of architecture. The interiors of most houses were designed around the fireplace as the dominant element. The exteriors also received a major element of their design from the mass and shape of the fireplace – or fireplaces in most larger houses – and the character of entire neighbour-hoods was established by the collection of buildings and their chimneys. One could not imagine colonial America, Dickens' London, or the Paris of the turn of the century without the fireplace and the chimney.

Central heating

With the development of central heating which followed the exploitation of coal, oil and gas as fuels a design phenomenon occurred which seems pathetic in retrospect. The mechanical engineer, the plumber and the sheet metal subcontractor mentality came to dominate our means of providing comfort in the interior of our buildings. The equipment they designed and the systems they fabricated were hidden in the bowels of the building or behind the surfaces of the walls as though we should be ashamed of them. Unfortunately like many creations of darkness they became homely, even ugly, and provided all the more reason why architects wished to conceal them from view. The modern house is serviced by oil or gas lines that sneak into the basement or utility room under the ground, converts the fuel in a burning unit completely automated and hence only seen by repairmen, and distributes the heat (or cool) quietly through ducts and pipes that are always a potential source of disaster if they fail. Even the grilles, diffusers, radiators etc are hidden from view or concealed in the woodwork. The thermostat remains the only visible part of the system to most people. My children are accustomed to hitting the thermostat if not enough heat is available, as though it contains all the secret ingredients within its tiny self. Except for a few of our modern architects who felt it was more honest to expose the ducts and pipes (and consequently earned the opprobrium 'brutalist') we have arrived at the age of the invisible heating and cooling system. Even the flue or chimney is difficult to find on most buildings of any size.

Now that we face clearly the end of the era of low-cost, easily available fossil fuels, it is· time to rethink in fundamental ways the response of buildings to climatic conditions and human comfort. It is time to think more about the natural sources of energy from the sun, from wind and from water. It is also time to rethink many of the basic principles of design.

There is a term I hear quite often in my visits to Great Britain. The words are, 'at the end of the day'. It is used sometimes to mean that the speaker is about to utter the last word on a subject. At other times it means the best alternative after all others have been considered. Still other speakers use the term to present what to them is the logical conclusion to an argument after all of the illogical conclusions of others have been listed. It is in this sense I wish to conclude my remarks.

CONCLUSION

Now that we are at the end of the era of low-cost, high volume fossil fuels, some will despair of our way of life which has been so dependent on these fossil fuels, because they can imagine no other. Architects can provide such imagination. We advocate an energy consciousness in our design of buildings, but we believe it has now the positive potential – not just a limitation.

Some of us see in any problem an opportunity as well. New inventions and innovations are emerging to cope, many of which are organised under the general subject of solar energy. A new thinking of our relationship to the sun and its bounteous and generous supply of heat, of light, and shade and shadow.

'At the end of the day' I see it all as the beginning of a *new* era of architectural design – the start of a bright, new tomorrow . . .

A time filled with challenges and opportunities.

A time to be responsive to a still wider set of human objectives in a world which becomes less contentious and hence a world more at harmony with natural resources and the forces of nature.

A time filled with as much commodity, firmness and delight as we have the courage and the imagination to create.

DISCUSSION

J. Harrington-Lynn (Department of the Environment)   In his presentation Mr Eberhard discussed proposed new standards. How are these standards to be complied with? Will calculations have to be checked for all new buildings?

J. P. Eberhard (AIA Research Corporation, Washington DC)   Our belief is that there has to be a family of tools and techniques. At the simplest level, that is, the dwelling unit single family residence, we believe that it should be possible to produce a manual of accepted practice. As long as the building that the home builder or developer is planning to use is approximately within the range of that manual of accepted practice, no further evaluation will be needed. The next level of sophistication which comprises the largest number of buildings in the USA, although not the largest group in real estate value, we believe are simple enough that adequate manual calculations should be possible. A further development which is already becoming available would be a simple programme for a programmable pocket calculator that can be purchased for 40 dollars. For large and complex buildings computer programmes will probably be necessary and a part of our responsibility in the US Government project is to evaluate the difference it would make, using one of the many computer programmes on the market.

My personal preference would be to remove thermal standards from the building codes and to set up a reporting system like the US internal revenue service reports. The building owner and/or his architect and engineer would file with the Government a report of the new building which has been designed and a statement of the energy budget for that building and the data that supports that energy budget. The Government would then decide how many and whether or not to audit such reports. For most purposes, and I assume the vast majority, those reports would simply be filed and that would be the end of it. The Government audit method would need to be well publicised even stated on the report form. Such a system would allow complete flexibility for the architect, engineer and building owner, provided the final building complied with the energy budget.

J. Harrington-Lynn   Obviously the approach to regulations in the USA is different from the UK. In the UK the regulations are interpreted by the local authorities and in the majority of cases all calculations are checked for compliance. There is, however, the possibility, under discussion at the moment, of self-certification by qualified designers.

J. P. Eberhard   I suspect you may have the same failure in the administration of the laws that we do, namely that the people who read the plans and the specifications in the building code office to decide whether you comply do not have either the time or sometimes the education to know what they are looking at. They develop what I call the 'stair-rail' syndrome, knowing how to check if there is a railing on the stairway. They know from past history that 20% of the time they can catch architects for having forgotten; this gives them credit for having done something positive and that is about all they look at when they get a plan.

J. Harrington-Lynn   It is perhaps worth noting in fairness to building control officers that in the UK over 80% of the buildings submitted for approval are not designed by qualified designers, and in many cases designers use the building control officers as consultants by submitting plans which do not conform to the regulations on the assumption that the officer will correct them.

R. Cullen (Architects Design Group)   Do we really need regulations at all? The RIBA and the architects in this country are getting bewildered by regulations. Many people have argued they are counter-productive in all kinds of sectors, and that really in the final analysis, if you want a good building, you have to have a good client, a good architect, a good engineer and a good team to produce that building and regulations just get in the way. Having been in the midst of regulations does Mr Eberhard share that view or does he still believe in regulations?

J. P. Eberhard   I share it in the advice I give. I spent five years with the National Bureau of Standards in Washington so I probably know the building code process from the inside out. I believe that Western society faces one of the crucial periods of challenge in its history. We in the United States most of all because we are the most profligate in our use of energy. There is unfortunately not much evidence in history that the democracies make easy and gradual adjustments to crises. They wait until the crisis is upon them and then try to respond. We must not wait for the real energy crisis.

How many people with professional responsibility can we convince of the seriousness of the situation and that it is encumbant upon them as professionals to provide immediately for the best possible solutions? How much power do they have with their clients to convince them? I am a realist and I accept that in the USA in the short period of time we have, market forces alone would not be sufficient to convince building owners and architects of the urgent need to be energy conscious in design now.

T. Smith (Steensen Varming Mulcahy & Partners)   At the cost of many millions of dollars, you are in the middle of a remarkably fast programme of information accumulation. In presentation three steps towards that were described. The last one of these was to invite designers to put forward the ultimate design, the effects of ultimate design in energy terms, and the result of those designs in terms of cost. Why are you doing the first two steps because, if there is the capability to produce designs which can encompass the best in energy conservation, these two exercises seem to be no more than a collection of statistical data, which, whilst extremely interesting, appear to serve little purpose. If you are capable of the third, why don't you go straight for it?

J. P. Eberhard   At a cost of 4.5 million dollars 168 firms are re-designing buildings that range from warehouses to hospitals, schools and churches (14 categories of buildings in all). These buildings statistically represent a cross-section of buildings in the USA. With that amount of effort invested it could be argued that as the target is technically feasible it should be used. The US Government still has that prerogative, but politically it would be difficult to lay down a national target because (a) not everyone is convinced we have an energy problem and (b) both professionals and building owners would feel they were being put upon by Government if they had to do what was maximum technically feasible as contrasted to having some latitude. The project will provide an indication of upper and lower limits – it will be for Government to select the compromise somewhere in between.

D. Fitzgerald (University of Leeds)   In presentation Mr Eberhard spoke of energy targets in a number of standard buildings in different climatic zones. I would be grateful for more information about these.

J. P. Eberhard   The first phase report* is available and this gives figures for energy buildings were designed to use in 1975 and 1976. Table 8.1 and fig 8.1 give this information and are reprinted from the report. Later a full report will be available and this will include illustrations of the buildings re-designed by the 168 firms commissioned by us.

G. Lesslie (Danfoss (London) Ltd)   Mr Eberhard has given information on new buildings. I imagine America shares our problem that the bulk of our energy problems lie in existing buildings. What is being done in the US to reduce energy use in existing commercial premises and dwellings?

J. P. Eberhard   The US Government is not proposing any general measure at the moment. If the energy bill gets through Congress in the form it has been drafted, subsidies will be available to non-profit institutions, hospitals, schools, colleges, to enable them to finance retrofit programmes. In the American Institute of Architects we are working on what I call redesign and in it I believe there is a great market opportunity. Every building designed within the last 20 years is probably now capable of being redesigned to make it more energy efficient. The process of doing that is architectural first, engineering second, and it must be in that order. To go to the owner occupier of an office building and talk about turning down lights and adjusting fan belts is starting at the wrong end. I say go to them as an architect and talk about their programme, what is going on in their building now. It almost

*Phase One/Base Data for the development of Energy Performance Standards for New Buildings – AIA Research Corporation, 1735 New York Avenue, Washington DC 20006. Published by US Department of Housing and Urban Development, Office of Policy Development and Research.

Table 8.1 Annual energy performance in 1000s BTU/sq ft

Building type		National	Region 1	Region 2	Region 3	Region 4	Region 5	Region 6	Region 7
					Heating & Cooling Degree Day Region				
Office	Mean	**64**	**65**	**76**	**65**	**61**	**51**	**50**	**64**
	20%	48	55	63	49	45	34	39	52
	80%	80	75	93	82	68	54	58	68
Elementary	Mean	**65**	**114**	**70**	**68**	**70**	**53**	**48**	**57**
	20%	47	*	54	51	53	39	38	*
	80%	79	*	84	86	80	65	55	*
Secondary	Mean	**114**	**Mean**	**68**	**55**	**51**	**37**	**41**	**34**
	20%	35	*	45	37	39	24	29	*
	80%	66	*	78	75	55	49	54	*
College/Univ	Mean	**52**	**77**	**70**	**46**	**59**	*	*	**83**
	20%	66	41	*	*	*	80%	80%	*
	80%	77	83	*	*	*	*	*	*
Hospital	Mean	**190**	*	**209**	**171**	**227**	**207**	**197**	**Mean**
	20%	113	*	*	*	*	*	*	*
	80%	231	*	*	*	*	*	*	*
Clinic	Mean	**69**	**Mean**	**72**	**71**	**65**	**61**	**59**	**59**
	20%	52	*	49	*	*	*	*	*
	80%	79	*	90	*	*	30	33	*
Assembly	Mean	**61**	**58**	**76**	**66**	**51**	**44**	**68**	**57**
	20%	40	*	46	53	42	45	102	*
	80%	69	*	103	75	58	89	104	104
Restaurant	Mean	**159**	**162**	**178**	**186**	**144**	**123**	**137**	**Mean**
	20%	105	114	111	106	102	155	104	157
	80%	210	184	221	232	153	40	154	*
Mercantile	Mean	**84**	**99**	**98**	**86**	**81**	**67**	**83**	**80**
	20%	62	72	67	69	72	86	55	*
	80%	99	120	134	97	91	*	100	*
Warehouse	Mean	**65**	**75**	**82**	**65**	**50**	**36**	**37**	**39**
	20%	38	*	61	48	*	*	*	*
	80%	83	*	99	80	*	*	*	*
Residential Non-Housekeeping	Mean	**95**	**99**	**84**	**94**	**125**	**90**	**93**	**106**
	20%	65	*	52	72	68	48	64	73
	80%	115	*	102	107	138	107	112	119
High Rise Apt	Mean	**49**	**53**	**53**	**52**	**53**	**Mean**	**29**	**Mean**
	20%	30	*	64	55	*	*	*	*
	80%	61	*	35	35	*	*	*	*
Multi Family Low Rise†	Mean	**43**	**58**	**55**	**41**	**31**	**27**	**22**	**32**
	20%	22	34	31	31	18	19	19	25
	80%	57	77	79	64	37	31	30	38
Single Family Attached†	Mean	**47**	**65**	**54**	**45**	**35**	**35**	**33**	**45**
	20%	26	39	28	26	19	32	24	31
	80%	63	81	71	68	54	37	31	53
Single Family Detached†	Mean	**69**	**104**	**73**	**61**	**52**	**43**	**38**	**58**
	20%	34	68	40	33	31	36	31	38
	80%	85	123	104	97	77	55	51	65
Mobile Homes†	Mean	**75**	**103**	**84**	**81**	**67**	**42**	**54**	**70**
	20%	53	66	55	54	49	29	52	53
	80%	96	132	102	98	83	48	83	97

* Sample size insufficient to calculate 20%, 80% and/or mean.
† These estimates include energy for space heating & cooling only. All other estimates include energy for space heating, cooling & lighting.

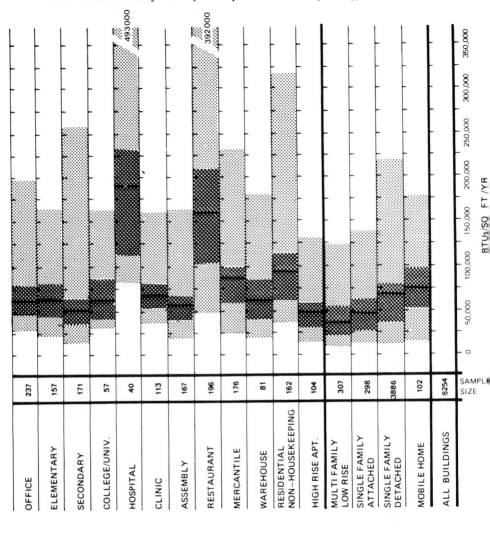

	SAMPLE SIZE
OFFICE	237
ELEMENTARY	157
SECONDARY	171
COLLEGE/UNIV.	57
HOSPITAL	40
CLINIC	113
ASSEMBLY	167
RESTAURANT	196
MERCANTILE	176
WAREHOUSE	81
RESIDENTIAL NON–HOUSEKEEPING	162
HIGH RISE APT.	104
MULTIFAMILY LOW RISE	307
SINGLE FAMILY ATTACHED	298
SINGLE FAMILY DETACHED	3886
MOBILE HOME	102
ALL BUILDINGS	5254

493000

392000

BTUs/SQ FT /YR

8.1 Energy performance distribution (all climatic regions)

NOTE: ESTIMATES INCLUDE ENERGY FOR SPACE HEATING, COOLING, LIGHTS, FANS & PUMPS.

NOTE: ESTIMATES INCLUDE ENERGY FOR SPACE HEATING & COOLING.

LEGEND

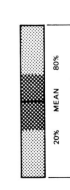

20%   MEAN   80%

assuredly is something different than the programme against which the building was originally designed. I would also ask them in developing this programme to be conscious of every decision about what is going on in this building, the kinds of spaces needed to house that activity, the relationship of those spaces to each other, that will have an impact on energy. We would have to decide together what design thermal and lighting conditions were really needed. In many cases clients have not even thought about this – just accepted what has been provided.

As an architect I would look at the building as a design problem, take the programme, the design requirements and see what I could do, given the existing building as a constraint but not a limit, to design a building which would house the activity and act as the primary modifier between external climate and what is going on inside. It is entirely possible that in a ten storey building, only nine floors would be needed, 10% of the energy could be saved just by replanning. It is entirely likely that in an office complex there are spaces that do not require to be air conditioned, eg dead file storage. If all such activities could be located on one floor the energy support required for that floor would be reduced. These are planning questions, something an architect is good at dealing with.

The next stage would be to look at possible use of non-fossil fuels (the sun, the wind) to add an additional measure of comfort to the activities in the building. After all that had been done, in most cases in the US it would be necessary to have additional mechanical and electrical equipment to produce the comfort conditions required in the building. Supplemental systems to handle the much lighter loads than would result from starting with a building and saying, let's retrofit it by cutting back from the existing sets of loads. That is the business opportunity in existing buildings in the US and there is probably the same opportunity in the UK for in both countries the inventory of existing buildings far exceeds the number of new buildings that will be built in the next few years.

9 Energy systems in the Royal Insurance Building

E. J. ANTHONY

INTRODUCTION*

In 1969 the Royal Insurance Co decided that they wished to have a UK Head Office in the traditional home of the company, Liverpool, where all the Head Office functions were already being performed in some seven separate offices. The client carried out an exhaustive evaluation of the functions to be performed in the building and defined the basic spaces required to accommodate these functions. A premium site on the Liverpool waterfront was eventually obtained which, while being excellent in location, had many restrictions associated with it which had impact on the shape and form of the building eventually designed.

In particular the need to fit into a pedestrian route system and the panoramic view over the Mersey were major factors in determining the orientation of the whole building and the location of different functional areas within the building.

The Royal Insurance Co's architect, J. D. Maxwell saw the need for an integrated design approach and set up at an early stage a team comprising Tripe & Wakeham, Architects, Bingham & Blades as structural engineers, Mercer & Miller as quantity surveyors and W. S. Atkins & Partners as building services consultants.

Because of the complex site conditions imposed by having a local authority car park accommodating some 1090 cars to be built beneath the Royal Insurance HO Building and to facilitate a continuous construction programme for the whole development, a negotiated contract arrangement was embarked upon with a Tyson/Haden Young Consortium which had essential experience and track record in building in Liverpool. The consultants and the nominated contract consortium had to undertake parallel working in design and construction to achieve the task required of them and by adopting an integrated design approach a building which clearly illustrates integrated environmental design was designed and constructed, occupation being commenced in the autumn of 1975.

The client wished to have the best environmental conditions which could be achieved compatible with reasonable cost and occupation and use of the

*The project was carried out using British units which are therefore used in the case study.

154

9.1 Royal Insurance Building, Liverpool, showing surrounding buildings.

building in accordance with an agreed programme related to the release of existing premises.

The basic accommodation plan showed that a gross floor area excluding car parking, of more than 57,000m² was required to accommodate some 1800 persons and that a large proportion of the space was to be open plan office. Additionally, staff dining facilities, a conference/lecture theatre, a medical suite, a training suite, two large computer suites, a bank and small shopping area were to be accommodated in the same building envelope.

SHAPE AND FORM

The factors which determined the basic shape and form of the building were:
- size of office units required for various departments and flexibility to accommodate changing operational patterns in the future
- relationship between departments
- site configuration
- environmental standards adopted
- economy in total cost.

It should be remembered that the conception and design of the building occurred considerably before the OPEC crisis of 1974, so that whilst energy considerations were important, running cost balances at that date were not so heavily tilted by energy and fuel costs.

However, it was apparent from the first considerations of the building that it would require to be fully air conditioned and therefore would be a large energy consuming building.

The viable development of the site inherently required a tall building of 12 storey height and since the building required air conditioning an economic air distribution system required air handling plant areas to be allocated at low level (LG), approximately half way up at +3 and above the top office floor at +11, thus keeping duct area lengths and air velocities within economic limits. Velocities were kept below 3500 ft/min as excessive velocities would result in power usage and the use of excessive silencing measures consuming additional energy. Four vertical cores were established for air duct accommodation and one core for pipe services between the refrigeration plant room at basement level, and the air handling plant areas.

DEVELOPMENT OF ENERGY SYSTEMS

The major factors affecting the energy consumption of the building were:
(a) environmental standards
(b) shape and form of envelope
(c) type of system. Designed to establish (a) in (b)
(d) possible heat recovery.

The principal environmental standards adopted were those applicable to the open plan offices which were 70° ± 2° F and 50% ± 5% relative humidity and a working plane lighting level of 1000 lux.

Whilst such simple standards may now be considered naive from an energy saving standpoint these were at that time accepted as progressive standards.

A whole series of studies were carried out prior to commencement of engineering systems design to optimise the shape and form of the building from an energy consumption aspect. If these studies were repeated today with much more sophisticated computer programmes being available, including our own ATCOOL, I doubt if in this aspect of envelope development any radically different solutions would arise. The building is a tiered pyramid of rectangular plan having approximately 15% fenestration area and since an appreciable amount of the total area is below ground, theoretical heat losses and gains are near optimum (fig 9.2).

Since the principal use of energy in the building was certain to be lighting and air conditioning, studies were made which evaluated the methods by which the high degree of illumination could be achieved with the lowest electrical energy input to produce 1000 lux at the working plane. At the same time as part of an integrated design exercise the possible means of meeting the cooling load in the building were evaluated and what was considered to be an optimised ceiling element selected, which performed air distribution and lighting functions.

The maximum heating load of the developed building fans being	3806 kW
and the maximum cooling load being	5270 kW

The ceiling element designed to meet the lighting, thermal and acoustic requirements is an egg crate ceiling. The basic ceiling module is 635 sq ft in area which is the basic structural module and consists of a high level

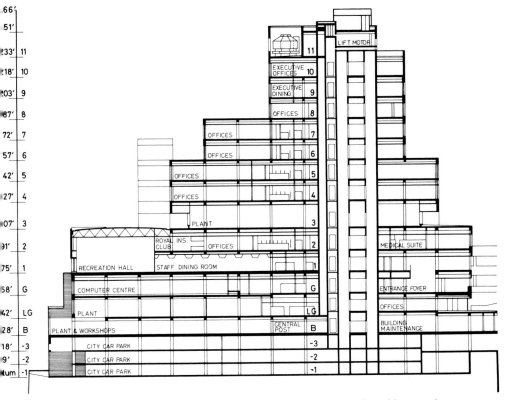

9.2 Royal Insurance Building, Liverpool; section showing tiered construction with approximately 15% fenestration area.

perforated steel plank ceiling with a fissured fibre board egg crate having 2 ft 1 in square cells suspended 10 in below the flat ceiling.

Lighting fittings are plain batten twin 6 ft 85 watt mounted directly on the flat ceiling above the egg crate. Air is supplied on a constant volume basis through flat plate diffusers which roll the air along the flat ceiling above the egg crate before diffusing to the area of the module through the egg crate. Extraction is through return grilles in the flat ceiling at 'moat' positions between egg crate modules. Each module has its own CV box and zone terminal heater battery so giving very flexible zone control to meet changes in usage and occupancy.

In the exercise to choose the generic type of air-conditioning system some seven systems were evaluated.

It may appear incongruous now that a variable volume system, shown as the lowest energy using system considered, was not adopted but in 1969 the use of such a system was ruled out largely because suitable proven system equipment was not readily available.

Full fresh air cooling facilities were also found not to be viable for inclusion throughout the building, such a facility being restricted to plants serving special areas such as the medical suite, cinema, gymnasium, kitchen and dining areas.

Economies can be achieved by including full fresh air facilities in design but it should not be assumed that such a provision will always be a major saver of energy and each case requires specific investigation in the early design stage before adoption.

Although a reduction in air quantity for cooling could have been achieved using air handling fittings the integrated ceiling described was found to be the best overall solution. Since the standard module was to be repeated over 400 times in the building a full scale mock-up was produced upon which air movement, lighting levels and acoustic properties were checked before finalisation of design. During this exercise the need for a fully ducted system was demonstrated as the use of a negative plenum ceiling void required 30% more air to effect full load cooling.

As the total cooling loads arising from occupancy, internal and external gains represented some 30% of the assessed energy requirement of the building it was apparent that considerable savings in running costs might be effected if the heating demands which coincide with cooling demand could be met by the heat of rejection of the refrigeration plant. In order to choose the method of meeting cooling and heating demand, five possible methods were evaluated.

The Liverpool Post and Echo development on the adjacent site to the Royal was proceeding ahead of the Royal project and it was known from commencement of design that a waste seepage of water from the Mersey Railway Tunnel was available for use in the development. The use of this supply therefore became an important factor in the selection of an optimum method of meeting heating and cooling loads. The options evaluated were:

Scheme 1
Conventional high temperature hot water oil fired boiler plant and a refrigeration plant of the packaged vapour compression type using cooling towers to dissipate the condenser heat.

Scheme 2
As Scheme 1 but using the river water direct for the condensing duties instead of the cooling tower equipment. The horsepower and operating costs of the compressors being reduced as lower condensing pressures result from the use of the lower river water temperature.

Scheme 3
Comprised a patented heat recovery heat pump system designed to fulfil all the heating and cooling requirements from one plant and one energy source. Full heating capacity can be obtained when the air conditioning refrigeration duties are zero, since river water can be the heating source as well as the dissipation sink for surplus heat of rejection from refrigeration plant.

Scheme 4

This scheme comprised a partial heat recovery vapour compression plant with supplementary oil fired water heating equipment, additional boiler plant being required during periods of low refrigeration load. The load patterns were such that the differences in the energy requirements could not be made up by the use of warm water storage. Cooling towers were used to dissipate surplus heat.

Scheme 5

This scheme was as Scheme 4 but the cooling tower system was replaced by river water cooling equipment.

To obtain the running costs of each option it was necessary to determine the daily load profiles for weekday and weekend working for each month of the year, the area under the profiles, integrated for the year giving the annual thermal demands. In the case of the schemes using heat recovery of the heat rejected by the condensers, a waste heat curve can be produced as a function of the evaporating duty. This curve when subtracted from the heating requirements curve, gives the additional heat required to be made up by the heat pump extracting from river water in Scheme 3 or provided by supplementary heating plant in Schemes 4 and 5.

Typical heat load curves are shown by figs 9.3, 9.4, 9.5 for the months of January, March and June. In January, additional heating is required over and above that obtained from the condensers. In June, there is sufficient heat available at all times to fulfil the heat demand and in March, supplementary heating is required but much of the waste heat can be utilised.

From the running costs and capital costs of each scheme the net present values of the five schemes were calculated. Scheme 3, the heat pump solution had the lowest NPV, Scheme 1 had the highest being 40% higher than Scheme 3. All capital and operating costs were based upon 1969–70 figures, and the experience if repeated today would be expected to show the heat pump to be even more favourable because of the lower cost of money and higher relative energy costs.

The prime energy input and comparative energy intake at the building curtilage was calculated for each of the five schemes. The heat pump scheme required 43% less prime energy than the then conventional systems represented by Scheme 1 (fig 9.6).

On the basis of the optimisation assessment the heat pump system was selected and the flow sheet for the system was developed as shown (fig 9.7).

The plant consists of Stal-Levin packaged water chilling plants with river water cooled, oil and condenser liquid sub-coolers.

The four 770 hp screw type compressors operate on R.22 refrigerant. The condenser circuit rejects heat to raise water from 85°F to a maximum of 105°F which is used as the heating supply to air handling plants and zone heaters, any heat which cannot be used for heating is rejected by plate heat

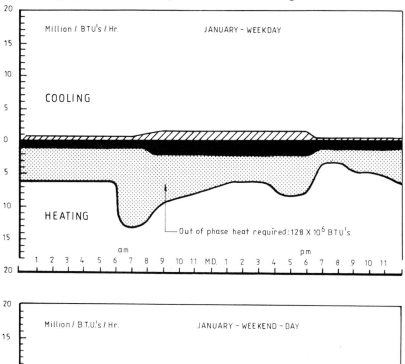

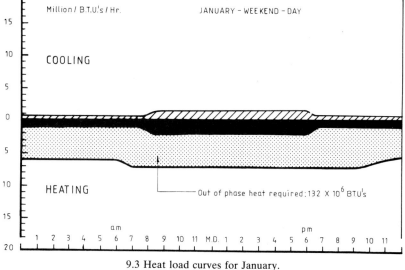

9.3 Heat load curves for January.

exchangers to water pumped from the Mersey Rail Tunnel. Plate heat exchangers were included in the design as being compact equipment easily maintained and suitable for working with small temperature differences and water such as the Mersey Tunnel water.

The evaporators take heat from the chilled water circuit reducing the chilled water temperature from 51°F to 41°F at full load condition. The chilled water is used to meet the cooling demands of the air handling plants.

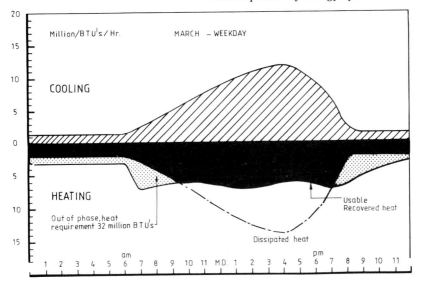

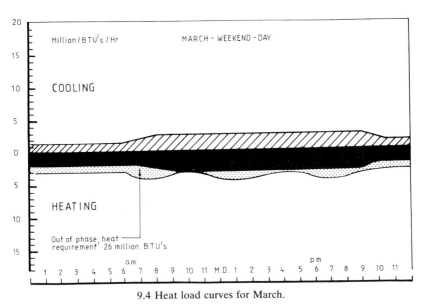

9.4 Heat load curves for March.

Plate heat exchangers can be connected across the external chilled water circuit to act as an artificial load when heat is required but no demand for cooling exists. Heat is then extracted from the Mersey Tunnel water and rejected by the condenser circuit to the air conditioning heating supply. To provide good control of the air handling units and chilled water temperatures and to prevent the undesirable characteristic of under cooling experienced by many air-conditioning plants, industrial type controls and

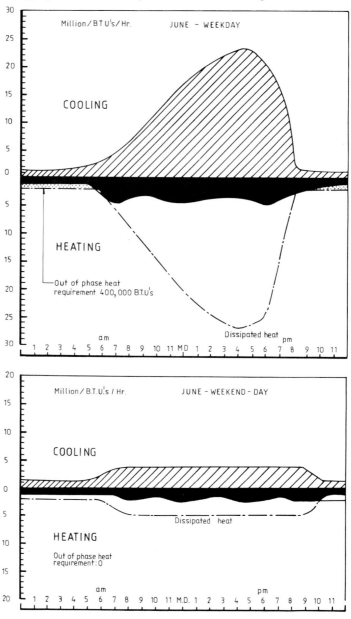

9.5 Heat load curves for June.

plant control techniques have been used. The water flow to the heating and cooling coils is controlled by straight through equal percentage characterised valves that give flow requirements relative to the duty and the heat exchange surface. This form of control requires less than proportional flow for any reduction of the coil duty. To maintain a constant flow in the

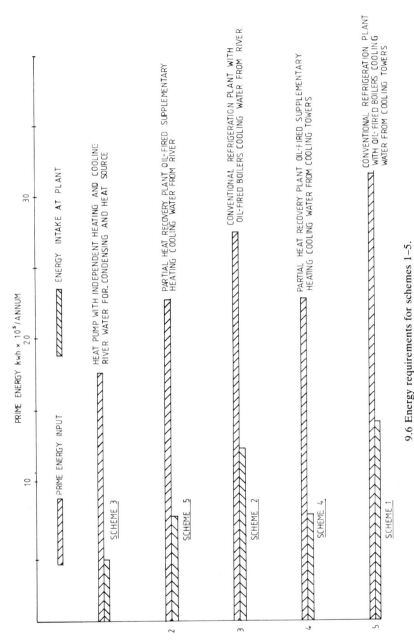

9.6 Energy requirements for schemes 1–5.

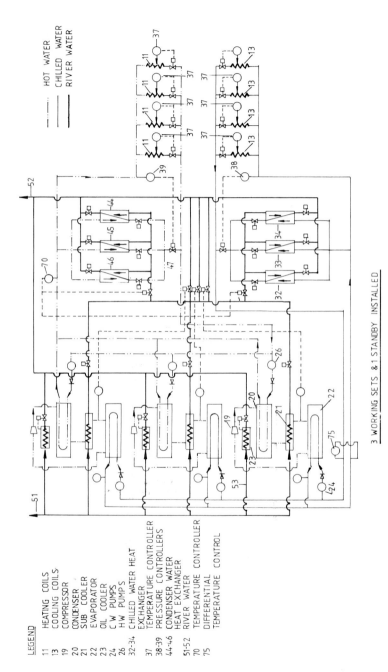

LEGEND

11	HEATING COILS
13	COOLING COILS
19	COMPRESSOR
20	CONDENSER
21	SUB COOLER
22	EVAPORATOR
23	OIL COOLER
24	C W PUMPS
26	H W PUMPS
32-34	CHILLED WATER HEAT EXCHANGER
37	TEMPERATURE CONTROLLER
38-39	PRESSURE CONTROLLERS
44-46	CONDENSER WATER HEAT EXCHANGER
51-52	RIVER WATER
70	TEMPERATURE CONTROLLER
75	DIFFERENTIAL TEMPERATURE CONTROL

---- HOT WATER
—— CHILLED WATER
—— RIVER WATER

3 WORKING SETS & 1 STANDBY INSTALLED

9.7 P and I diagram for heat pumps.

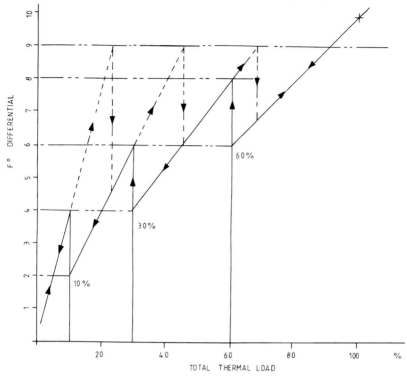

9.8 Diagram showing centralised heating and cooling plant and machine plant loading.

evaporator and condenser circuits a by-pass valve is inserted between the flow and return to control at a constant discharge pressure and flow to each chilled water set while working. This system gives variable water flow in the coils within the air handling plants and constant flow in the chillers. A differential temperature measuring device controls the number of machines operating for a given duty, while the capacity of each machine is independently controlled. A diagram (fig 9.8) plotting the total thermal demand and the differential temperature shows how the control is effected. A temperature controller measures the difference between chilled water flow and return 10°F at maximum load. When the differential falls to 6°F ie 60% load when four chilled water pumps are operating the lowest duty selection machine and its associated chilled and condenser water pumps are switched off.

At this point the temperature will rise to about 8°F as a result of only three chilled water pumps running and the fourth machine will only cut back in if a 9°F difference is reached. Further fall in duty operates stops at 4, 2 and 1°F temperature difference when the load on the first duty selection machine will be about 10% of the unit, full load rating, ie 2½% of full design system load. When the system load increases, the compressors will be brought into operation in sequence when the temperature difference reaches 9°F. This difference may decrease when further chilled water pumps and units are brought into operation. If the load is increasing and the differential is maintained, further loading of units takes place at timed intervals until a temperature difference of less than 9°F is maintained or until all units are running.

The sub-cooler being an energy saving piece of equipment, should be incorporated in any high condensing temperature system if practicable.

The evaluation made excludes some considerations which arose during development of the total design. In order to provide cooling to the building power generation unit it was necessary to provide some cooling tower capacity, so that the installation of cooling towers has not been avoided completely. Boiler plant is required for the production of domestic hot water as it is not economically possible to extend the heat recovery and heat pump plant to provide water at the temperature levels suitable for domestic usage. However, the addition of these factors to the optimisation assessment would not alter the order of merit of the various options considered.

It is the Royal Insurance Co policy to have sufficient power generation capacity available to permit the sustained operation of the building under power failure conditions, not merely to have a standby capacity sufficient to satisfy safety and emergency needs.

Thus 2.5 mW of diesel generating capacity is installed in a building which to date has a maximum demand of 3.54 mW. This demand is considerably below that predicted at initial design stage, principally because of economies effected in the lighting. By the time the commissioning of the installation commenced the lighting level was reduced to approximately 700 lux by taking out of use some 30% of the lighting installed in the building.

The lighting design includes window splay reveal lighting to all windows in the open plan areas and wall washing lighting to all internal core walls. This lighting is separately switched and can be programmed out from the Delta 2000. External floodlighting amounting to some 230 kW was designed but for energy economy reasons has not been fitted. Many areas such as lift lobbies and foyers had much decor lighting using down lighter fittings but these are not used in the normal operation of the building as satisfactory lighting intensities are achieved by the general area illumination system.

Lighting tubes were chosen principally for efficiency and designed into the integrated ceiling component to give the required illumination levels, the open plan area illumination being achieved at a dissipation of 32.6 watts/m^2.

Additionally computer loads were initially assessed to include the use of new equipment which has currently not been installed. This reduction in dissipation within the building has meant a considerable alteration to the self-heating character of the building and the adjustment of systems to optimise on running costs is an operation which is still going on and requires the steady building of operating history to enable sensible final setting up and operation routines to be effected.

In order to achieve as much energy saving as possible from the reductions in internal gains, all plant volumes were reassessed and assessments made of the theoretical savings to be made by reducing the temperature set point by 4°F during heat demand periods and raising back to 72°F during demand for cooling periods. Running of the energy programme showed theoretical annual savings of 22% in heating energy requirement and 50% in energy requirement for cooling. However, it must be said that attempts to date to effect these savings have given rise to an increase in complaints of discomfort largely from female staff.

An essential part of the engineering services of a complex building is the Building Services Control and Monitoring System. The system installed is a Honeywell Delta 2000 having some 3000 control and monitoring points associated with the air conditioning, ventilation, lighting energy consumption and safety of the building. The control Centre contains the central processor with its programmable memory, one illuminated digital display for all commands, replies and alarms with a keyboard to address all the connected points and a free standing printer to make hard copies of displayed data and to prepare operating summaries for building management analysis. A slide projector gives, automatically, graphic diagrams of any part of the building system selected for surveillance. Plant start optimisation and trend logging are aspects particularly intended to effect energy saving.

10 Case study of building design response to increasing energy costs

L. S. GINSLER

INTRODUCTION

Responding to the well publicised scarcity and increasing cost of energy in the world and particularly in the American Continent, the Provincial Government of Ontario has directed that certain of their new institutional buildings being planned shall be designed not only to present-day recognised high standards of energy economy, but in addition shall have mechanical plants incorporating storage and solar heat absorption.

These latter provisions are well beyond normal commercial practice at the present time and on the basis of present fossil fuel prices the savings resulting from their use do not justify the capital expenditure. However, fuel oil and natural gas, the present major fuels for building heating in the Province, are expected to double in cost every four to five years in the foreseeable future and established domestic reserves of them have exhaustion projections of the order of a few decades.

The intention, therefore, is to promote the development of a technology of building internal climate production that will:

- use no petroleum fuels at all
- use electricity most efficiently and economically
- make the maximum use of such solar absorption as is available in the various climatic conditions which exist in the Province.

The particular building considered in this chapter, a Provincial Courthouse for the Metropolitan Toronto Borough of North York, is one of the projects for which this special design authorisation has been made.

The branch of the Provincial Government responsible for the implementation of this design work is the Ministry of Government Services whose staff have prepared the brief outlining the building requirements in detail. The actual design is being carried out by Page & Steele, a prominent Toronto architectural firm, who have appointed Rybka, Smith and Ginsler Ltd for the mechanical design. By direction of the Ministry we are to work in close collaboration with it, represented by G. Kellner, the chief of its mechanical staff and K. Linton, mechanical engineer on that staff.

BACKGROUND

Energy use in Canada is high, on a per capita basis probably the highest in the world. Two reasons are advanced for this. The first factor is the climate which dictates building heating for a large part of the year, and secondly, the universal and extensive use of motor vehicles in the large, thinly populated country.

Energy resources are considerable, and at the moment, probably sufficient for national requirements, but geographic considerations obtrude, namely fuel transportation costs.

Thus the Atlantic provinces have coal but import oil, Quebec imports coal and oil and exports hydro-electric power. Ontario imports coal, receives oil and gas over the pipe from the west and hydro-electric power from Quebec and exports uranium fuel, and electricity to the United States. The West, Prairies and Pacific Coast have oil, natural gas, coal, nuclear fuels and export all four.

So Canada is a large importer of fuel principally oil, but to some degree in the East Central Region (Quebec and Ontario) coal.

Electricity in the province is generated and distributed by a provincial publicly owned commission, Hydro Electric Power Commission of Ontario. Province wide generation (1975) was as follows:

Water power	57%
Coal fired	22
Nuclear	16
Petroleum product (gas)	5
	100%

Water power in the province is now largely all developed. The coal fired stations were built in the 50s and 60s pending the slower nuclear development, which now appears to be on its way to becoming the principal source. The small percentage of natural gas used is burned in the summer by the coal-fired stations in big cities to mitigate pollution.

Prior to about 1935 coal was almost universally used for building heating and it was imported mainly from the US. Subsequently it has been displaced practically completely by, at first fuel oil, and then by natural gas after the pipeline from the west was built (in the 1960s) and to a small degree by electricity, the use of which was promoted by the Commission at about the same time.

This has been a considerable revolution. Coal handling and storage areas in all the Great Lakes ports together with the ships that carried the coal have disappeared, there are no more coal trucks on the streets and municipal ash collection services no longer operate in the cities. The iron and steel industry and the Power Commission's thermal plants, which have enormous requirements, are practically the only users remaining.

As mentioned above, the Power Commission have promoted the direct use of electricity for building heating, and some, relatively very small, proportion of newer buildings have been so equipped. Of these, some have had reasonably acceptable operating costs where the highest standards of insulation were used together with extreme moderation in fenestration. In other cases where such considerations were ignored, operating costs have been disastrously high. However, changing over existing buildings to electric heating, to say nothing of the provision of required additional generation and distribution capacity would be prohibitively expensive. Basically, the direct use of such high grade and expensive energy as electricity for widespread building heating must be considered inefficient and uneconomical.

Heat reclaiming refrigeration (heat pump) is feasible where internal heat gains in the building are sufficient to provide for building heat losses, and many such installations are in operation. These provide heat, raised in temperature to a usable condition at the expenditure of electric energy for reversed refrigeration in amount of approximately 20% of the total made available by this process. This represents therefore, a reasonably efficient use of electric energy. Use of such systems have mainly been in office buildings where lighting loads are usually high enough to make them feasible. However, they only work when the lights are on, and the display of downtown office towers in Toronto burning lights night and day and over weekends to heat the building by the lights and running refrigeration as well, is beginning to strike observers, urged by their Power Commission to 'turn off that light', as being inefficient and wasteful. The answer to this is, of course, storage.

CLIMATE

The Toronto climate is generally considered mild in winter in comparison with that of the major part of the country. Design outdoor temperatures are −5°F winter and 92°F dry bulb, 75°F wet bulb in summer. Over the year, 6850 degree days F below 65°F is generally considered the proper figure. It is one of the most southerly cities in the country – roughly the same latitude as the French Riviera. However, it receives much of its winter weather from the north-west and its summer weather from west and south-west. The Great Lakes on which it is located have a moderating effect on temperature, but also are responsible for the considerable winter cloud cover.

As noted above, summer dry and wet bulb temperatures can be high, the cooling season extends from early May to mid October, and heating can be required from early October to mid May.

REGULATION

One evidence of the concern about energy is the proposed issue of an 'Energy' supplement to the National Building Code of Canada, the draft of which has been circulated for consideration by the professions concerned. It regulates minimum thermal resistance rates for walls, roofs, floors, windows and such elements, maximum percentages of fenestration and the like. It also is concerned with types and arrangements of air-conditioning, heating and ventilating systems and requires heat reclamation where feasible. The intention is that the Code, when finalised, will apply to new buildings of most categories throughout the country. Anticipating the first issue of this Code by several months the provincial government of Ontario make similar requirements of an energy conserving nature in the design of new buildings they commission with the architectural and engineering profession. In addition to design requirements they also ask for a maximum calculated energy consumption rate, designated 'energy budget', expressed in kWh/ft^2 of floor area per annum, for all energy used in the building including lighting, heating, air conditioning, plumbing, lifts and other building services. This calculation is a computer operation, based for heating and air conditioning on a typical year, ie one that most closely approaches an 'average' year, for which a weather tape is available giving hourly dry bulb and dew point temperatures and cloud cover throughout the year.

BUILDING REQUIREMENTS

The building we are concerned with is a courthouse for the Metropolitan Toronto Borough of North York, the design of which has been commissioned by the Ontario government. Energy conserving factors to be provided in the design are as follows:

1. U Factors   Walls 0.08 Btu/ft^2 h °F
                  Roofs 0.06 Btu/ft^2 h °F
                  Glazing 0.35 Btu/ft^2 h °F
                  and 0.25 shading factor.
2. Window areas not more than 40% of floor to ceiling wall area, applying to any 15 ft long module of the building. However, balance temperature of the building shall be no higher than 25% of the net difference between indoor and outdoor design temperatures.
3. Ventilation rate – office occupancies – .08 CFM per net usable sq ft.
4. Lighting levels prescribed are considerably reduced over what has been customary in the recent past as follows:
   General and public areas   120– 180lux
   Circulation area in office   240–360 lux
   Normal office work   400–600 lux
   Office work – prolonged visual   600–900 lux
   Reflectivity of interior surfaces, prescribed as follows:
   Ceilings   80%
   Walls   50%
   Floor   20%
5. Interior design temperatures were prescribed as 72°F winter, 76°F summer.

6. A heat reclaiming refrigeration plant with storage and sun assistance is included.
7. Energy budget for the building is to be not more than 14.7 kWh/ft²/yr.

BUILDING

The building has basement, ground floor and three upper stories, and is 187,000 sq ft gross in total floor area. Facilities included are as follows:

Basement	Drivers training, plant rooms, building services and maintenance office and shops, unexcavated area.
Ground floor	main lobby, seven traffic courts, administration areas, court reporters' offices.
Second floor	8 criminal courts, 2 storeys high, 9 judges' offices, prosecutors' office, 3 small claims courts.
Third floor	smaller because of second floor court rooms, holding cells, interview rooms, court offices and the like.
Fourth floor	3 large jury court rooms extending through roof, 4 family courts, judges offices and the like.

Exposures are as follows:

Total exterior wall (including windows)	67,178 sq. ft.
Total roof area	53,033 sq. ft.
Total soffit (overhung floors)	2,929 sq. ft.
Windows and entrance doors	8,711 sq. ft.
Design population	1,200 people

BUILDING LOADS

The following are building loads as presently established. As noted above, these may vary somewhat when final computations have been made.

1. Building sensible cooling load at summer design condition (93°F DB)	1700 MBH
2. Design air quantity for cooling at 20°F difference at the coil (17°F diff at the outlet)	90,000 CFM
3. Fresh air quantity 0.08 CFM per net usable sq ft	11,000 CFM
4. Cooling load, including fresh air	250 tons
5. Winter design transmission heat losses at −5°F	860 MBH
6. Winter fresh air load	775 MBH
Winter total heating load	1635 MBH

7. Winter internal cooling load, minimum
   including lighting and people, but
   without sun gains   1580 MBH
8. *Balance temperatures*
   (a) Transmission losses as against lighting only $= -38°F$
   (b) Transmission losses and minimum fresh air
       against lighting only $= +14°F$
   (c) Transmission losses and fresh air against
       lights and people $= + 2°F$

WORK CYCLE OF THE BUILDING

The intended work cycle of the building throughout the week and the year is
an important factor in the calculation of requirements, particularly where
storage of heating and cooling source is involved. In this case specific data
has been established as follows:

- the whole building will operate on a Monday through Friday week
- the ground floor, where traffic courts are located, will operate 15 hours
  per day, since as a matter of convenience to the public such courts operate
  through the day and in the evenings
- the remainder will operate 9 to 5 and our figures allow for 10 hours per
  day
- the building will be closed on holidays
- the operating year will be a full 52 weeks.

MAIN PLANT SYSTEMS

The main plant consists of the equipment outlined below. It is indicated in
the heating/cooling flow diagrams (figs 10.1–10.8) which for clarity show
only one refrigeration machine with its pumps, chiller and condensers, and
one cooling tower.

(a) Two centrifugal refrigeration machines, each with water chiller
    (evaporator) and two full capacity condensers, one for winter heating
    and one for summer heat rejection to the cooling towers.
(b) Circulating pumps of constant capacity characteristic are provided in
    conjunction with each machine, for chilled water, heating condenser
    water, and heat rejection condenser water, a total of six pumps for the
    two machines.
(c) A cooling tower is provided for the heat rejection of each machine, a
    total of two.
(d) A solar heat absorption system having solar collection panels on the
    roof of the building, together with heat exchangers in the basement, and
    dual circulation pumps (total four), circulating glycol solution and water
    at constant flow.
(e) Two large storage tanks are provided:

(i) solar heat storage tank, which operates in conjunction with the solar absorption system on a seasonal basis.

(ii) diurnal storage tank, which operates in conjunction with the refrigeration plant on a daily basis.

(f) Water circulation systems, each equipped with one variable and one constant speed pump, to give variable flow, each set being of double the capacity of the individual pumps at the refrigeration machines, for the heating and cooling circuits to the coils in the air handling systems.

(g) A transfer pumping plant is provided for transferring water between the two storage tanks. This pump is reversible.

(h) Finally, a small electric actuated heating boiler is provided for emergency and standby use.

The refrigeration plant is sized so that the total cooling capacity of the two machines is 90% of the design maximum load. The intention is that, by reason of the storage, only one machine will be required to be operated at any time, and the other will serve as standby. It was considered necessary to provide the two machines, since the plant will run all year. In many instances in Canada, where refrigeration is operated only in the summer, satisfactory operation has been obtained with one machine for many years, but this mode of operation involves regular and extensive inspection, maintenance and repair in the winter season when the machine is not required. At the moment, the individual machine size is set at 110 tons.

The solar absorption panels are located on the roof each consisting of a sheet metal tub heavily insulated in the back, having a copper sheet on top of the insulation with copper tubing brazed thereon, the whole being faced with double glazing. Panel size is 4 ft wide by 8 ft high and they are arranged in rows facing south, inclined at an angle of 53° above the horizontal (10° greater than the latitude).

Tests have indicated that these panels, properly located, will have a minimum net annual heat absorption in the Toronto climate of 22,000 BTU per sq ft per annum, and the area being provided is 3,000 ft. Certain more highly developed and expensive panels are claimed to have considerably better performance. As noted above ethylene glycol and water solution will be circulated in this system to allow pump shutdown under control of a photo cell, at night and when sun absorption is not effective due to cloud cover. The use of anti-freeze solution is considered essential in this climate, where so-called 'non-freeze' systems using water have frozen up.

The storage tanks are of long narrow configuration, with cross barriers and down-pipes at alternate ends to reduce mixing of warm and cold water to a minimum. They are located in the basement plant room within the building.

Sizes are: solar tank   70,000 imp gallons
            diurnal tank   130,000 imp gallons

Preliminary indications are that the prescribed performance will be exceeded, and that the annual energy consumption will be only 10 to 12 kWh/ft² including lighting consumption of about 5 kWh/ft². In addition electrical efficiency is enhanced, and cost of energy reduced by the reduction of demand and load factor improvement allowed by storage.

The Ministry has directed that the choice and design of a Building Management System to operate the lighting, heating and air-conditioning systems in the building shall be deferred. The system for this will have to deal with condition of storage in the two tanks established by probes located throughout the two tanks, the various cycles conforming to the cycles required throughout the year and possibly the expected weather, all addition to the normal requirement for occupancy programme and existing weather conditions.

AIR SYSTEMS

Interior areas

For the interior areas of the building, ie those not having external exposure two air-conditioning systems are provided – one for the ground floor and the other for the remainder of the building. The reasons for this division are the longer hours of operation required for the ground floor area and for true energy economy it is very important that the maximum shutdown times be taken advantage of.

These systems will basically be high pressure variable air volume type having 90% shutdown control on the variable air volume elements. They will be for cooling only and will be provided with 100% fresh air availability to allow free cooling when required. In addition, sprayed cooling coil humidification will be provided. The sprays will mainly be used for evaporative cooling of the supply air, at times when this process is required.

The systems as outlined above are used for all interior areas other than the court rooms, where during operating hours the loads will be reasonably constant.

The courts are also all served for supply air by these systems, and are regarded as interior whether or not they have some exterior exposure. However, variations are provided as follows:

- Air distribution to the room is 'all air induction', having boxes in the ceiling space with primary air supplied by the interior systems and inducing the flow of return air from the ceiling space through the boxes to the room for temperature control. Thus the circulation rate in the room will be reasonably constant, this feature being considered desirable in these large ceilinged, variably occupied rooms.
- Wall and ceiling heat losses to the rooms will be eliminated by cavity wall heating on exterior walls, and roof space heating by unit heater where roof exposure occurs.

Exterior areas

These are the areas, apart from court room areas, around the perimeter of the building and are normally 10–15 ft deep. The systems are divided for ground floor and the remainder of the building as are the interior systems and for the same reason. After consideration of several alternatives, the following arrangement was chosen:

- Cooling is provided by two high pressure variable air volume air conditioning systems similar to those provided for the general areas of the interior, each having the additional capacity to handle transmission and solar heat gains, as well as internal load.
- Transmission heat losses will be provided for by a warm water heating circulating system having fan coil units at the floor, under-the-window and wall mounted. Thermostats for the units, which will have automatic night set-back, will control an automatic modulating valve on each unit, with an end switch controlling the fan motor, starting and stopping it in conformity to the valve position.

This arrangement provides the following:

(i)   Sequential cooling-heating control, the heating only coming on when cooling is at minimum in the zone, and thus simultaneous heating-cooling with resulting needless energy expenditure is obviated.

(ii) The air-conditioning system is inherently the most economical type, and can be entirely shut down outside of working hours, winter and summer.

(iii) The heating system can be independently operated in off hours at small power consumption, and heat energy consumption only as required to satisfy the requirement.

(iv) Heat will be available at near shutdown (10% flow condition) of the air conditioning system to mitigate over-cooling under these conditions.

(v)   The fan coils are the most economical means of doing the heating using the low temperature heating water.

As against these advantages the following disadvantages are accepted:

(i) The fan coil units take up minor space in the occupied areas.

(ii) Servicing, mainly filter changing, will be required in the occupied areas.

(iii) Unit selection will have to be carefully made on a noise basis.

(iv) A water circulating system above the level of the storage tank surface will be necessary. This involves some extra pumping power and complication.

Reclaim from extract air

For maximum heat reclamation, required in the depth of winter, cooling coils are provided in extract air discharges to atmosphere as follows:

- on toilet extract system
- on air-conditioning extract outlet, size based on minimum fresh air flow condition.

THE DIAGRAMS

The diagrams show the equipment and installations comprising the heating and cooling plant, together with the interconnecting piping and control

LEGEND TO HEATING - COOLING SCHEMATIC

O	OPEN		
C	CLOSED	CONTROL VALVES	
M	MODULATING		
CS	CONSTANT SPEED PUMP		
VS	VARIABLE SPEED PUMP		
	3 WAY MODULATING MIXING VALVE		
	STRAIGHT THROUGH ON-OFF VALVE		
	CIRCULATING PUMP(S)		
H.E.-1	WATER TO ANTIFREEZE SOLUTION HEAT EXCHANGER		
	CHECK VALVE		
	SYSTEM COOLING COIL		
	HEAT RECLAIM COIL		
	IMMERSION CONTROL THERMOSTAT WITH MAIN AIR SUPPLY		
C-S-1 / C-W-1	DOUBLE BUNDLE CONDENSER S-SUMMER USE W-WINTER USE		
E-1	EVAPORATOR NO.1		
TANK TS	AUXILIARY HEATING TANK SUPPLY WATER SERVICE		
TANK TR	AUXILIARY HEATING TANK RETURN WATER SERVICE		

P-1	SUMMER CONDENSER WATER PUMP-CS
P-2	WINTER CONDENSER WATER PUMP-CS
P-3	CHILLED WATER PUMP-CS
P-4	MAIN HEATING SYSTEM PUMP-CS
P-5	MAIN HEATING SYSTEM PUMP-VS
P-6	MAIN CHILLED WATER SYSTEM PUMP-CS
P-7	MAIN CHILLED WATER SYSTEM PUMP-VS
P-8	AUXILIARY HEATING PUMP #1-CS
P-9	AUXILIARY HEATING PUMP #2-CS
P-10	REVERAIBLE TRANSFER PUMP-CS
P-11	SOLAR HEATING SYSTEM WATER CIRC.PUMP#1-CS
P-12	SOLAR HEATING SYSTEM WATER CIRC.PUMP#2-CS
P-13	SOLAR HEATING SYSTEM #1-CS ANTIFREEZE CIRC.PUMP #1-CS
P-14	SOLAR HEATING SYSTEM ANTIFREEZE CIRC.PUMP#2-CS
P-15	AUXILIARY BOILER PUMP SET-CS

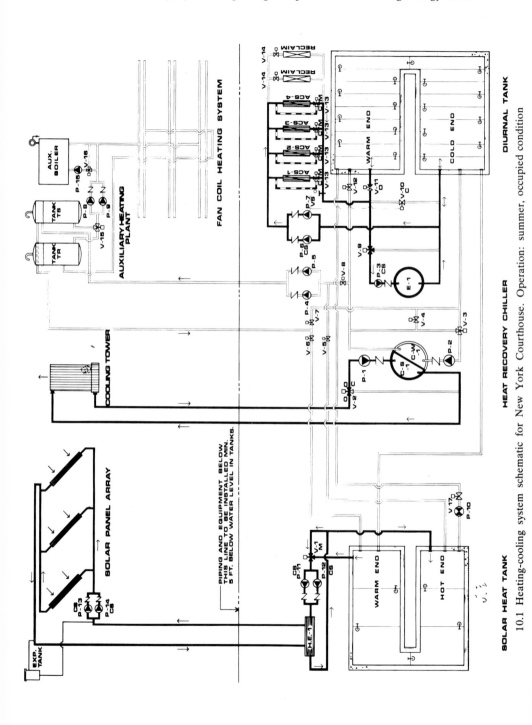

10.1 Heating-cooling system schematic for New York Courthouse. Operation: summer, occupied condition

valves involved in changing the flow patterns to accomplish the various cycles of operation that will be required in different seasons of the year.

Water temperatures
The water temperatures (at design) envisaged are as follows:

Refrigeration machine	Summer		Winter	
	On	Off	On	Off
Chilled water, °F	62	42	62	47
Condensed water, °F	85	95	85	105

Diurnal tank

	Summer	Winter
Warm end, °F	57–62	100–105
Cold end, °F	42	47

Solar tank

	Cold	Warm	Hot
°F	45–47	85–105	180

VARIOUS CYCLES

Various cycles that will be operated throughout the year in conformity with the weather and storage status in the two tanks are illustrated in figs 10.1 through 10.8, and are discussed below:

Figure 10.1
This diagram illustrates operation in late summer and early autumn, occupied hours, and indoor design condition will be 76°F.

Solar absorption system
This system will operate during daylight hours, occupied and unoccupied. Ethylene glycol solution is circulated through the solar panels and heat exchanger HE-1 at constant volume by pumps P.13 and 14. Storage water is circulated through the heat exchanger by pumps (constant volume) P.11 and 12. As the water in this circuit becomes 180°F Valve V-1 acts to admit warm water from warm end of the solar tank and discharge hot water to hot end. By the end of the summer, the solar tank will be completely hot (180°F).

Heat requirement
During the summer, heat requirement of the building is nil, and the heating system will not operate.

Heat rejection
All heat rejection from the refrigeration plant will be by the cooling tower and condenser CS-1 and pump P-1. Valve V-2 will operate to prevent too

low condenser water temperature in cooler or dryer weather, to ensure stable refrigeration machine operation. The water flow in this condenser circuit will be greater than in the circuit through CW-1, as required by the lower temperature rise of this water, as noted above.

Cooling cycle

All four air-conditioning systems will operate on minimum fresh air. Chilled water requirement will be greater than the output of one refrigeration machine, and so the two pumps P6 and P7 will handle a volume of water greater than the capacity of the machine pump P-3. The excess will be drawn from the cold end of the diurnal tank and discharge to the warm end of same. This process is 'taking from store'.

Unoccupied hours

The solar absorption system will operate in daylight hours.

The four air-conditioning systems will be shut down throughout unoccupied hours, lights will be off and human load will not exist. Pumps P-6 and 7 will be shut down. The refrigeration plant and pump P-3 will operate, drawing water from the warm end of the diurnal tank and discharging chilled water to the cold end 42°F. This process is 'storing' and will continue until sufficient store is established for the next day.

Figure 10.2

This diagram illustrates the cycle of operation in October and early November, when cooler weather begins to exist.

The solar absorption system will continue to operate in daylight hours when sun conditions justify it to maintain the solar tank as hot as possible, compensating for use of the hot water and tank losses.

In occupied hours, indoor design temperature will revert to winter conditioning 72°F, and the following processes will function.

The four air-conditioning systems will operate as much as possible on free and evaporative cooling. When the required air discharge temperatures cannot be achieved by these means, the chilled water cycle will start up, providing the required chilled water to the cooling coils, first from the cold end of the diurnal tank as shown in the diagram, and then, if more capacity is required by the starting up of the refrigeration machine. This operation of the machine will depend on the store condition of the diurnal tank, which will be maintained in a 'balanced' condition throughout this period to take care of the warm 'Indian Summer' days that occur occasionally most years throughout this season. When refrigeration is run during this period, heat rejection will be to the tower, to obviate undue refrigeration use in maintaining storage.

Heating requirement is minimal in this season, most of the transmission going to reduce cooling load. It will be required mainly for warm up in the morning from the night setback condition and the remainder for

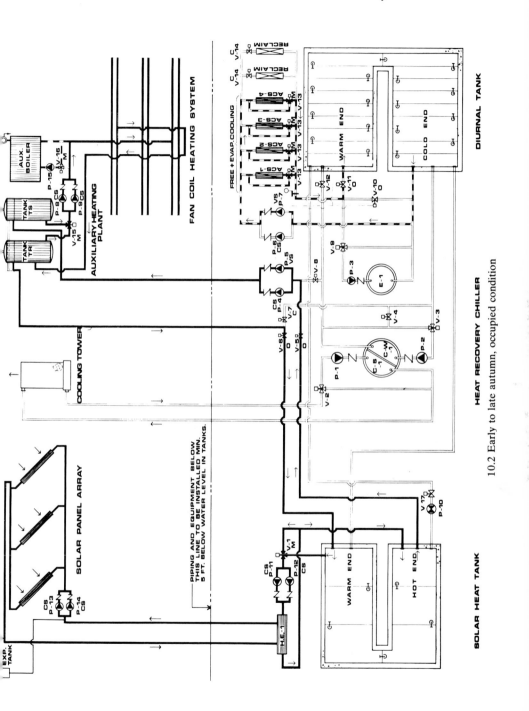

10.2 Early to late autumn, occupied condition

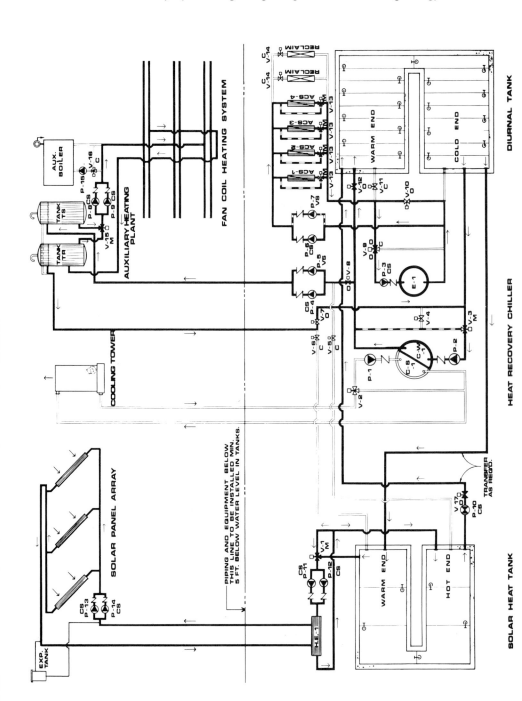

compensating during the day for the overcooling effect of low load shut down in variable air volume systems.

Such heating as is required will be provided using hot water from the solar tank directly. The diagram indicates the mechanics of the 'above water level' heating system. Solar tank water will be supplied at 180°F to the 'supply tank' TS, and will be returned by the heating system to the return tank TR at 85°F from which it will overflow back to the cooler end of the solar tank. The heating water will be mixed to 100°F from tanks TS and TR through valve V-15 and pumped through the heating circuit by pumps P-8 and P-9 back to tank TR.

Unoccupied hours

All air conditioning and refrigeration will be shutdown. The heating system will operate. The night setback temperature (60°F) will operate. Due to heat storage of the building, this setback, and relative mild weather, little heat will be required. Such as is required will be available.

Figure 10.3

This diagram illustrates conditions and cycles of operation during late autumn and early winter, the latter half of November and all December, occupied times.

The solar absorption system continues to operate and early in this period the solar tank will drop to 'warm' condition (100°F or less) and the mixing valve control (V-1) will reset to 100°F. Use of this water for heating directly will cease. The transfer of warm water from the solar tank to the warm end of the diurnal will start and be carried out as chilled water for its replacement is available at the cold end of the diurnal tank. Sun absorption will operate as available in daylight hours.

Heat rejection from the refrigeration machine by the cooling tower will cease.

The cooling load will be less than the output of one machine, and refrigeration capacity control will be by the cooling load, and no use of chilled water from store will be required for cooling duties. The air systems will be operating on minimum fresh air to provide a cooling load.

Heating load will be increasing, and on occasion may reach as high as 75% of design. However, the heat will be less than that available through the winter condenser from the cooling load, and since at this season the heating water will be re-set down because of the milder outside temperature, some admixture of chilled water from the cold end of the diurnal tank, and equivalent discharge of heated water to the warm end of this tank will occur.

Figure 10.4

This diagram illustrates operation of the plant in unoccupied hours for as long as stored heat is available in the diurnal tank from either warm water in the solar tank or that produced during the daytime by operation of the refrigeration plant.

Although the diagram does not so indicate, some operation of the solar absorption system will be possible, in the manner shown in fig 10.3, during the daylight hours of weekends and holidays.

The refrigeration machine operates to cool the stored warm water in the diurnal tank under control of the heating requirement which is always less than the output of one machine, discharging chilled water to the cold end of this tank. Heat is thus provided through its condenser to the heating system. The auxiliary boiler only comes on if supply water to the heating system drops below the required temperature.

Transfer warm water from the solar tank in exchange for chilled water continues until the solar tank is fully cold.

Figure 10.5

This figure illustrates conditions and cycles during occupied hours in mid winter (January and February).

This situation is similar to that portrayed in fig 10.3 except that we now need all possible heat, and the reclaim coils in the exhaust discharge from air conditioning and toilet exhaust systems come into operation. Further the stored heat in solar tank only amounts to the daily small increment from the solar absorption system, and the diurnal tank is also running out.

The building will support itself during occupied hours down to the design temperature ($-5°F$) but the heat available for store during the day is minimal at design temperature but greater in normal winter temperature for this period.

Figure 10.6

This diagram illustrates how the plant would operate in unoccupied times, during a very cold spell after the storage tanks are exhausted, which might occur in the first half of March. This condition would be quite unusual.

The auxiliary boiler would provide for the night and weekend heating load under these conditions.

Figure 10.7

This diagram illustrates operation in occupied hours early and mid spring, specifically mid March, through April.

It is the same as fig 10.3 except that transfer between solar tank and diurnal tank is reversed and chilled water is being exchanged from the solar to the cold end of the diurnal tank, being replaced with warm water from the warm end of the diurnal tank to the cooler end of the solar tank.

In unoccupied hours, during the early part of this period, required heat will be provided as illustrated in fig 10.4, providing heat from the refrigeration machine winter condenser and adding chilled water to store.

In the latter part of this period, no heat will be required night or day.

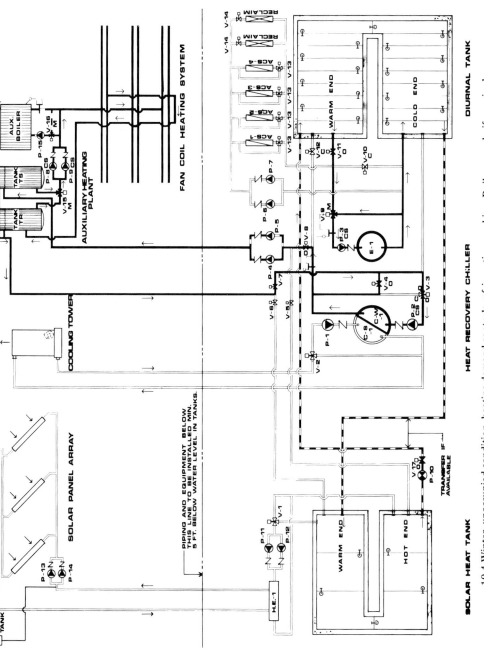

10.4 Winter, unoccupied condition, heating demand controls refrigeration machine. Boiler on only if required.

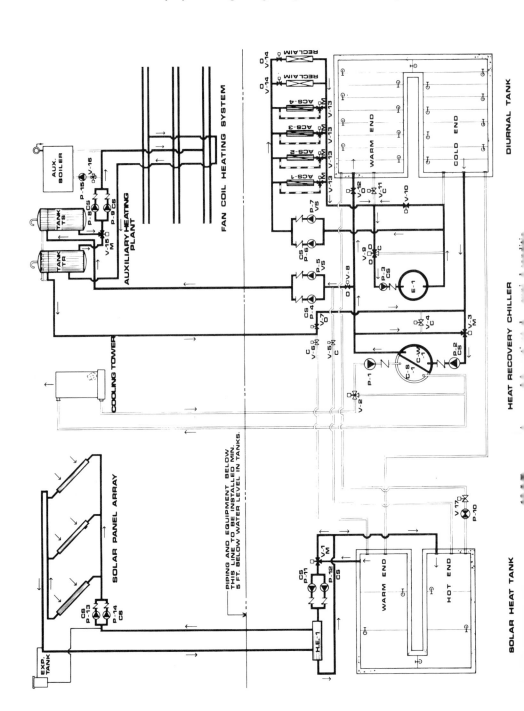

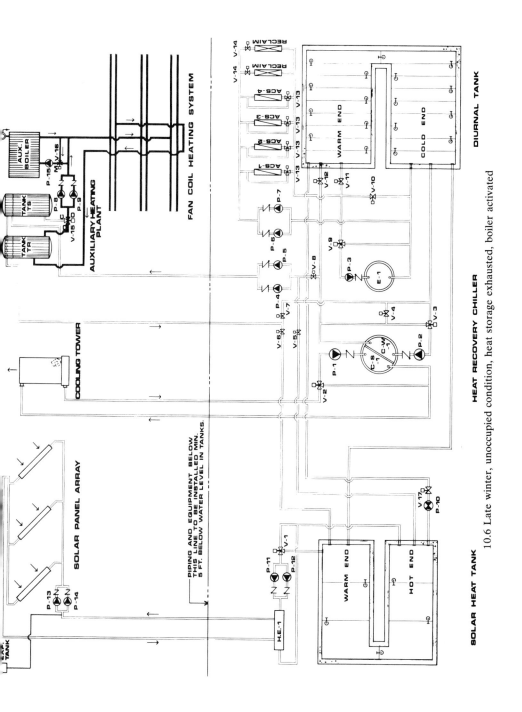

10.6 Late winter, unoccupied condition, heat storage exhausted, boiler activated

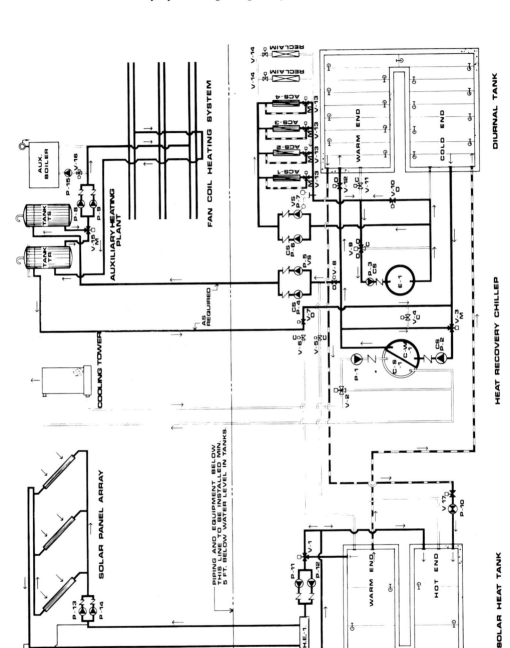

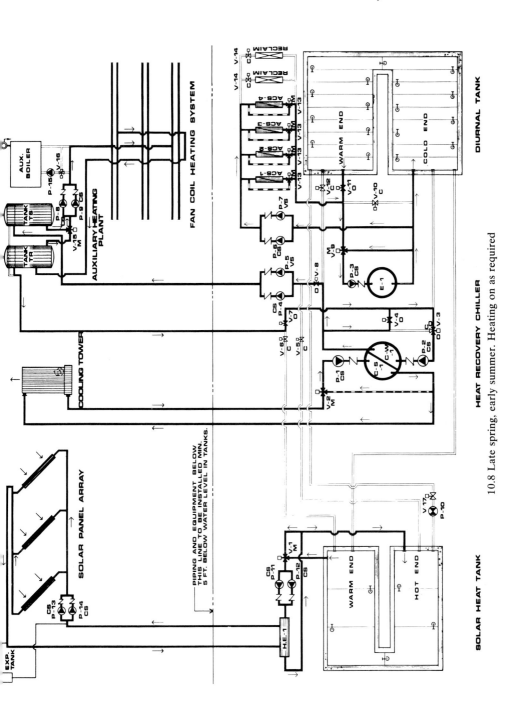

10.8 Late spring, early summer. Heating on as required

Figure 10.8

This diagram illustrates operations, late spring and early summer, occupied hours specifically May to mid June.

The solar tank is completely warm, and the sun absorption process is well established, and large store of chilled water hereby created in the diurnal tank.

Little if any building heat is required and excess rejected heat from refrigeration begins to be discharged from the cooling tower.

Cooling load is less than the output of the machine and thus its operation in occupied hours will increase the chilled water storage in the diurnal tank.

In unoccupied hours the plant will be completely shutdown.

DISCUSSION

L. Longworth (University of Manchester)   What a breath of fresh air it is to hear someone has investigated ways of managing with a refrigeration plant smaller than the maximum load. I certainly do not believe there is widespread application of working out a maximum instantaneous refrigeration load then cutting it in half making up the difference by thermal storage. I am quite sure we should emulate the Canadians in this and quickly – after all we already do something similar for domestic hot water.

Mr Ginsler said that he has found the total fan pressure provided in an installation based on calculations of duct resistance is frequently too high. He advised investigating this before starting to balance the system. I wonder where he feels this error arises, presumably in the ducting system and particularly in evaluation of local resistances. The source of data on the resistance of branches and bends in duct systems is very difficult to find. At Manchester we have initiated a new look at the resistance of ducts and particularly at branches for this very reason.

L. S. Ginsler (Rybka Smith & Ginsler Ltd, Toronto)   Having previously accepted fans with an available static pressure which is too low we now add a little more pressure into the specification just to be sure. After all, we know we can always slow the fan down but it is sometimes very embarrassing to speed it up. We also, in designing sheet metal ductwork, are including historic safety factors which no longer need to be applied. For instance, architects have been educated to the point where they now do provide a legitimate space for ductwork with resultant reduction in the overall pressure drops through the system. No longer do we use ducts that are 750 mm × 150 mm let alone the duct I saw once which was 144 in × 1 in (3600 mm × 25 mm). I really think it is just a general overall improvement caused by increasing experience of designers and contractors and available products.

There has been a quite dramatic new development in North America of low pressure variable volume terminal devices. Historically manufacturers stipulated a minimum of ¾ in of water static pressure (1.9 mbar) then realistically we found this not to be accurate and we had to use a minimum of 1¼ in. (3.1 mbar). There are new devices on the market with pressure drops as low as ⅓ to ½ in (0.8–1.3 mbar) and the trend is to slowly work our way back to using lower pressures.

E. J. Anthony (W. S. Atkins & Partners)   The design and contract procedure very often influence how equipment is sized in the first place. Negotiated contracts and various design procedures which arise for reasons outside basic engineering tend to force the engineer to leave himself a margin.

L. J. Wild (George Wimpey & Co Ltd)   Whilst congratulating Mr Ginsler on aiming towards low energy usage (I hope this country will introduce similar legislation to that in Canada) I have a nagging worry that the introduction of the services hardware, the building elements, insulation elements and the increased cost of structure, together with the greater area required to house the hardware, mortgages some of the future energy savings and may in fact accelerate and worsen the usage situation.

L. S. Ginsler   In Toronto, Canada, insulation is automatic. The only question is the degree of insulation. To our amazement we found that if the thickness of insulation was doubled the capital cost was only marginally increased because a good deal of the cost was related to labour rather than the material itself.

Introduction of energy conservation features produces an almost embarrassing reduction in mechanical and electrical requirement. We have reached the point where for instance our loads are so low that the air volumes are insufficient to provide adequate air movement. For this reason all air induction systems are seriously being considered. I think they are not entirely valid unless you get into the ultimate kind of complexity I presented in my paper, which was not intended to be a valid economic exercise, rather a research programme. On the other hand in the newspaper plants we have designed the money saved on refrigeration and heating equipment has justified the overall cost. The only real significant premium is the storage tanks and it is building the tanks out of concrete that really creates complexities in design. Unfortunately it is much cheaper than steel.

A. Howland (Pfizer Ltd)   In presenting his paper Mr Ginsler indicated he had experimented with reducing the condenser pressures of refrigeration plant in an effort to reduce overall power consumption but for various reasons decided against it. To do this it would probably be necessary to run the cooling tower fans appreciably longer, if they are under thermostat control, or to have more of them to achieve the lower cooling water temperatures. I would welcome comment on the relative balance of power between bringing condenser pressures down and increasing fan running time.

L. S. Ginsler   It is necessary to make an individual assessment on every job. For instance, using an induced draught type of tower where the horsepower is relatively light, added power requirement for the fans was more than offset by the saving in the improved thermal efficiency of the refrigeration plant. When we had the Baltimore tower using twice the horsepower for forced draught the saving went in the opposite direction (negative). We are also considering the possibility of using variable volume on cooling towers using inlet dampers.

E. J. Perry (W. S. Atkins & Partners)   I would like to comment on the last question and answer. There is a great deal of energy wasted by the use of expansion valves fitted to refrigeration plant and use of constant temperature controls across the cooling towers. A number of plant designers and suppliers should modify their designs of equipment and plant to minimise this loss of energy and the loss of operating costs that the owners of plant incur due to this type of control. The loss of energy due to expansion valves and constant temperature condensing systems have been investigated in depth*. The savings in energy, and as a result operating costs that can be achieved by allowing condensing temperature to follow diurnal changes of temperature can be considerable. The design of the plant may have to be modified so that these changes can be achieved in practice. With flooded evaporation systems this modification is easy, but with expansion valve systems greater consideration should be given to changing the system design.

D. Arnold (Troup Bywaters & Anders)   To laymen solar collectors are perhaps the only evidence of energy conservation. Mr Ginsler indicated that the solar collectors shown on his illustrations are to be removed or exchanged. Toronto is on the same latitude as the French Riviera but I am sure the temperature is not the same. Could Mr Ginsler have justified solar collectors without the very tight brief provided?

L. S. Ginsler   The popular press have given so much publicity to solar energy that we have to defend not using solar panels on practically every job. For example in a senior citizen building in Toronto we were asked to generate domestic hot water using roof solar panels. Our assessment was that the payback period was over 100 years.

Our situation is a little unique because we carry the penalty of having to use an anti-freeze solution. Because of toxicity problems we must go through two heat exchangers for heating domestic hot water. I have seen solar panels used very effectively in the Middle East, but in our climate and under our conditions, no way.

J. Harrington-Lynn (Department of the Environment)   Mr Ginsler said that one of the requirements they were working to was that internal heat gains must at least equal heat losses. Is it intended to make this requirement a legal one in Canada?

L. S. Ginsler   It is not the law yet but it certainly is a requirement on any Government building – including some we are doing now in Sudbury where the temperatures go to −40°C (−40°F).

*Energy and Refrigeration Systems, Design, Maintenance and Operation, Institute of Refrigeration (UK), 5 May 1977.

J. Harrington-Lynn Mr Wild raised the question of the economics of energy conservation measures. In the UK any legislation proposed for energy conservation must be justified on economic grounds, for the country as a whole, ie the cost of the measure must equal the cost of the energy saved over the life of the measure.

This is particularly true of the measures Dr Cunningham referred to in the Introduction. For example the proposals for thermal insulation of non-domestic buildings due to be laid before Parliament shortly* were arrived at after a detailed survey and analysis of 150–200 buildings, which had been submitted to local authorities for Building Regulation approval in the last few years.

Estimates of the extra cost of insulating these buildings to a higher standard and the energy saved by their insulation were made and shown to be economic.

R. J. Bentley (The Electricity Council) I was particularly interested in Mr. Ginsler's co-operation with architects. I wonder if he has been engaged in work where the building envelope is used as a thermal modifier, ie the appropriate building materials are selected to get the right ratio of thermal transmittance and heat storage so that energy savings are achieved both on cooling load and, with intermittency of occupation, on the heating load.

L. S. Ginsler Consulting engineers really have never had that kind of power. We do take into account thermal storage factors in our selection of plant. However, I have as yet to receive a commission where we have had the privilege of exerting an influence on the construction materials. The structure development and architectural development are normally long resolved before we have an opportunity to check what influence it has. This may change and perhaps I am influenced by the predominant use in Canada of concrete structures which are inherently massive.

E. J. Anthony There are available a number of computer programmes which enable buildings to be modelled which take into account the storage characteristics of the building. The extent to which they have been used to advantage in the design of buildings is a matter for question. They seem to have been used so far to investigate existing buildings to discover why they are not operating satisfactorily rather than as a means of setting down parameters for the design of new buildings.

B. Z. Gillinson (Gillinson Barnet & Partners) I am surprised to hear some of these remarks. I talk to a lot of fellow architects. I talk to a lot of building services engineers and work with them and to my mind it is common practice to discuss these matters with our consultants at the brief stage. We discuss the thermal characteristics of the building in the total concept and I do not think we are unusual in doing this. Tom Smith described the job we did together (chapter 4) and our two teams were talking about these subjects from day one.

E. J. Anthony You are right in saying that we all talk and have done for a

*New regulations for non-domestic buildings became operative in June 1979.

long time about the use of envelopes as a thermal modifier. However, real analysis taking into account the equally important thermal mass of the internal contents as well as the external envelope, is rarely done in my experience. When such analyses are carried out it is because there has been some failure of the design which has arisen because of characteristics revealed after the design is completed and the installation does not work.

L. S. Ginsler   Certainly the envelope is given very serious consideration and also the finishes and standards of lighting. However, I cannot think of a single building where storage factors ever even entered into the discussion.

D. Fitzgerald (University of Leeds)   I would like to recommend the Admittance Methods for calculating the non-steady state thermal behaviour of buildings. It is a manual method developed in the UK and easy to use in the early stages when utilisation of computer techniques might not be possible. Mr. Anthony is correct that there are many programmes to evaluate non-steady state behaviour of buildings but by the time all the data needed by a computer programme have been assembled it may well be too late. Using the admittance method it is possible to do a quick calculation using the rough data obtained at the first meeting of the design team, ready for the second meeting next day. The method gives results of engineering accuracy, comparable with the accuracy of the thermal data of building materials. This method is in the CIBS Guide, but lacks a definite exposition with explanations and examples. I teach the method to students of architectural engineering in the University of Leeds. At the moment I have the great privilege of teaching it to a group of American students of architectural engineering from Pennsylvania State University, currently studying at Leeds.

11 Capital and running costs of air-conditioning systems

R. B. WATSON

INTRODUCTION

As society has become more sophisticated so the demand for a more closely conditioned environment has increased. This demand has arisen at a time when greater attention is being paid to building economics. The need for cost control has become increasingly apparent and the use of elemental cost analysis and cost planning has led to a reduction in disproportionate expenditure on the individual elements of a building.

The establishment of elemental cost planning some 20 years ago has led to the use of what are now recognised as more refined techniques for monitoring the cost of building elements and recently these techniques have come to be used effectively for engineering services elements.

It appears that the trend in the cost of building elements has tended to reduce in proportion while the cost of engineering services elements has increased. This is partially due to the introduction of more sophisticated fire protection services and communication installations but it is the ventilation and air-conditioning elements which have contributed most to the increase in engineering services expenditure.

The purpose of this chapter is not to examine the justification for air conditioning but to examine techniques to monitor capital costs during the design process and to establish running costs so that the value of air conditioning can be determined with more precision than hitherto.

As air conditioning is introduced into the design of a building, the effect in terms of cost can only be established if a sound basis for cost forecasting is available. The decision can then be made as to whether or not air conditioning will become economically viable in the context of a particular project.

COST PLANNING

The basic principles of cost planning depend upon the division of the building and its services into elements. The cost per m² is then calculated for each element. Initially this cost is obtained from the cost of the equivalent element of a similar building, and an adjustment made using element ratios or quantity factors (1).

Definitions to ensure that each element contains the same components

195

have already been established under the Standard Form of Cost Analysis produced by the Building Cost Information Service (9). Element ratios enable adjustment to be made for buildings containing different quantities of a particular element. Element ratios or quantity factors suitable for engineering services are now being more widely adopted (2) and a suitable factor for ventilation or air conditioning would be m³ per second/m². This allows inter alia an adjustment to be made which reflects the cubic capacity of the building and the number of air changes.

Costs per m² can therefore be adjusted to reflect differences of quality and quantity for different buildings and the use of cost indices to eliminate the effects of inflation will normally enable the use of historic data to be used with impunity for cost planning purposes.

THE COST PER M² METHOD

When the air-conditioning element is expressed in terms of cost related to area, it is found that there are many factors which, individually, can influence the result dramatically. The primary factors will relate to the effects of the application of the quantity factor together with those factors which include the shape of the building (narrow/deep plan), the type of glazing (single/double, solar resistant, etc), the proportion of glazed area in relation to the total area of outer walls and finally the magnitude of the area to be air conditioned.

Capital costs relating to air-conditioning systems in both narrow and deep buildings were presented in tabular form on the basis of cost per square metre in a previous publication (3). Similarly different costs were tabled to establish the differentiation between buildings of 20% and 50% glazed area (3). In many respects these costs appeared to enable relatively precise forecasting to be carried out where previously little or no information had been available. These can be seen by reference to table 11.1.

It was stated however that the selection of double glazing to minimise energy use (4) and reduce running costs (3) proved difficult to assess properly in terms of effect on the capital cost of the air-conditioning system. As the costs in table 11.1 relate to buildings with a wide spectrum of differing areas and types of air-conditioning systems, it is therefore worth examining these in closer detail before examining the possible effect of double glazing.

There are in addition other factors which affect the cost per m² and the development of cost planning techniques to date therefore does not accurately reflect these factors.

THE USE OF QUANTITY FACTORS

Having examined the effect of building shape and glazed area on the air-conditioning cost, the next factor which influences the cost per m² is the area of the building itself. In this respect air-conditioning costs should be

Table 11.1 Capital cost of air conditioning system in deep buildings
Costs as at January 1978

(a) 50% Glazing

Gross building area m²	2 000	3 500	5 000	7 500	10 000	15 000	20 000	25 000	30 000
Description of system	**Capital cost of system £/m²**								
VAV with two-pipe induction to perimeter	57.38	51.49	50.73	48.26	48.45	47.12	45.41	45.13	44.84
VAV with two-pipe fan coil to perimeter	57.00	50.92	50.07	47.41	47.79	46.36	44.84	44.46	44.18

(b) 20% Glazing

Gross building area m²	2 000	3 500	5 000	7 500	10 000	15 000	20 000	25 000	30 000
Description of system	**Capital cost of system £/m²**								
VAV with VAV dual duct at perimeter	51.21	45.98	43.23	42.94	41.42	41.80	40.66	39.81	39.62
VAV with perimeter reheat	46.27	41.71	39.52	39.62	38.48	38.76	37.62	36.77	36.67
Remote four-pipe fan coil	43.89	38.95	36.48	36.10	34.20	34.49	33.35	32.49	32.30

after Mitchell and Leary:
Environmental Quality and its Cost

related to the treated area or the area of the building actually air conditioned.

Table 11.2 sets out some recent examples of analyses of air-conditioned buildings arranged in ascending order of cost per m². The costs relate to treated area but upon examination area alone appears to have only an indirect effect on the cost and even the selection of the type of air-conditioning system appears to have only a limited effect.

These major factors cannot therefore be the only ones to influence the cost per m² and the introduction of the appropriate quantity factor, which in this case would be the cost per m³ per sec, should lead to a further significant adjustment. The quantity factor in this case is derived by dividing the total cost of the air-conditioning system by the total cubic metres per second rating of all the fan outputs. This quantity factor appears to work in relation to straightforward supply and extract ventilation systems but it is not sufficiently accurate a method when applied to air conditioning and is only useful for initial cost budgeting when insufficient information is available to implement a more sophisticated method. The main reason for this is that, whereas the numbers of air changes and total cubic capacity of the building are reflected in such a quantity factor, the additional requirements of cooling and humidity control for air conditioning lead to distortion of the rate per m².

Further factors therefore require to be considered and these will include the many different concepts of air-conditioning systems developed for use in this country in the earlier part of this decade, the choice of fuel and type of heat source plant, the choice of cooling plant, and the methods of heat and air distribution. Factors of minor consideration relate to water treatment, methods of automatic control, sound attenuation requirements, standby requirements, need for fire protection and the disposition of the main plant.

To accommodate these additional factors more sophisticated and detailed techniques appear to be required.

Table 11.2   Results of analysis of air conditioning installations
Costs as at January 1978

Ref	Type & location of building	£/m² Treated area	Treated area m²	Remarks
1	Library; Cardiff	50.08	6 050	LV central supply & central re-heat
2	Library; Aberystwyth	59.18	6 400	LV central supply: central re-heat
3	Offices; Middlesex	62.32	5 754	2 pipe fan coil: LV system
4	Office block; Lancashire	70.00	7 028	2 pipe induction system
5	Offices; Woking	72.80	5 400	VAV system
6	Office block; Manchester	73.09	9 800	4 pipe fan coil: LV system
7	Office complex; Berkshire	77.68	33 084	VAV system
		84.21	6 025	LV central supply: central re-heat
8	Offices; London	79.52	3 920	HV fan coil: re-heat system
9	Offices; London	86.49	3 813	VAV with terminal re-heat system
10	Medical school and dental hospital; Yorkshire	87.22	34 816	LV central supply terminal re-heat
11	Offices; London	87.85	10 691	VAV system
12	Offices; London	88.61	13 940	Terminal re-heat HV system
13	Bank and offices; Manchester	96.10	2 280	4 pipe fan coil: LV system
14	Superior offices; London	102.16	8 000	4 pipe fan coil: LV system
15	Teaching laboratory; Midlands	118.82	5 000	LV central supply terminal re-heat
16	Offices; Portsmouth	139.60	8 922	VAV system
17	Offices, shops, flats; Surbiton	147.23	564	LV central supply terminal re-heat
18	Office building; Swindon	161.14	6 300	LV 2 and 4 pipe fan coil, local air supply
19	Offices & shops; London	172.24	1 734	VAV system
20	Offices; Whyteleafe, Surrey	172.52	2 054	4 pipe fan coil; LV system

THE COMPONENT BREAKDOWN METHOD

As air-conditioning systems continue to increase in sophistication and develop in complexity the sole use of a cost per m² will prove to be increasingly fallible. The additional requirement of identifying the effect of energy conservation in terms of capital cost must lead to separate examination of the main components, or sub-elements, which together contribute to the total cost of air conditioning. If these can be identified separately this should enable the immediate financial effect of a particular design proposition to be ascertained.

Table 11.3 Component breakdown method – variable air volume system

See table 11.5 Note ref		Office block of 6 000 m²		Office block of 15 000 m²	
		Cost of element £	Cost of element per m² floor area £	Cost of element £	Cost of element per m² floor area £
	Boilers				
A	Plant and instruments	11 700	1.95	18 000	1.20
B	Flue	4 800	0.80	6 750	0.45
C	*Water treatment*	4 800	0.80	8 250	0.55
D	*Gas installation*	1 800	0.30	2 250	0.15
	Space heating				
E	Distribution pipework	12 000	2.00	23 250	1.55
F	Convectors and/or radiators	2 400	0.40	6 000	0.40
H	*Heating to batteries*	24 600	4.10	41 250	2.75
J	*Chilled water to batteries*	18 300	3.05	30 750	2.05
	Condenser cooling water				
K	Distribution pipework	9 000	1.50	14 250	0.95
	Cooling plant				
L	Chillers	36 000	6.00	57 000	3.80
M	Cooling towers	4 200	0.70	6 000	0.40
N	*Automatic controls*	26 700	4.45	50 250	3.35
	Ductwork				
P	Supply	166 800	27.80	366 000	24.40
P	Extract	36 000	6.00	79 500	5.30
	Air-conditioning plant				
Q	Heating batteries	11 100	1.85	18 000	1.20
Q	Humidifiers/cooling batteries	20 400	3.40	33 000	2.20
R	Fans and filters	32 100	5.35	51 750	3.45
S	Sound attenuation	16 800	2.80	36 000	2.40
T	*Fire detection*	10 500	1.75	17 250	1.15
V	*Electrical work in connection*	12 000	2.00	21 750	1.45
		£462 000	£77.00	£887 250	£59.15

Table 11.4 Component breakdown method – induction system

See table 11.5 Note ref		Office block of 6000 m²		Office block of 15 000 m²	
		Cost of element £	Cost of element per m² floor area £	Cost of element £	Cost of element per m² floor area £
	Boilers				
A	Plant and instruments	11 700	1.95	18 000	1.20
B	Flue	4 800	0.80	6 750	0.45
C	Water treatment	4 800	0.80	8 250	0.55
D	Gas installation	1 800	0.30	2 250	0.15
	Space heating				
E	Distribution pipework	12 000	2.00	23 250	1.55
F	Convectors and/or radiators	2 400	0.40	6 000	0.40
G	Induction units	41 000	6.85	102 750	6.85
H	Heating to batteries	46 800	7.80	71 250	4.75
J	Chilled water to batteries	58 200	9.70	101 250	6.75
	Condenser cooling water				
K	Distribution pipework	8 700	1.45	12 000	0.80
	Cooling plant				
L	Chillers	27 900	4.65	45 750	3.05
M	Cooling towers	3 900	0.65	5 250	0.35
N	Automatic controls	33 300	5.55	58 500	3.90
	Ductwork				
P	Supply	48 300	8.05	83 250	5.55
P	Extract	24 300	4.05	50 250	3.35
	Air-conditioning plant				
Q	Heating batteries	8 400	1.40	14 250	0.95
Q	Humidifiers/cooling batteries	14 400	2.40	21 000	1.40
R	Fans and filters	21 000	3.50	36 000	2.40
S	Sound attenuation	11 400	1.90	27 750	1.85
T	Fire detection	8 100	1.35	14 250	0.95
V	Electrical work in connection	10 500	1.75	16 500	1.10
		£403 800	£67.30	£724 500	£48.30

Table 11.5 Brief specification notes relating to tables 11.3 and 11.4

Ref

Boilers

A   Plant and instruments: three gas-fired boilers each of approximately 250 and 580 kW capacity for the two buildings respectively; together with burners, pumps, direct-mounted instruments, feed and expansion tanks. Full standby capacities are not included, pumps are in duplicate.

B   Flue: mild steel insulated in boiler house, mild steel internal lining to vertical builders stack.

C   *Water treatment*: chemical dosage equipment.

D   *Gas installation*: pipework internal to building, meter, solenoid valves.

Space heating

E   Distribution pipework: pipework from boilers to terminal equipment, all valves, fittings and supports, insulation.

F   Convectors and/or radiators: panel radiators, or natural convectors in circulation areas and staircases.

G   *Induction units*: high velocity units suitable for four-pipe system utilizing ducted fresh air.

H   *Heating to air heater batteries*: distribution pipework to batteries and including valves, fittings and supports, insulation.

J   *Chilled water to batteries and induction units*: distribution pipework including valves, fittings and supports, insulation.

K   *Condenser cooling water*: distribution pipework, valves, fittings and supports, insulation.

Cooling plant

L   Chillers: centrifugal chiller units of approximately 190 and 470 tons total capacity for the two buildings respectively, including mountings and supports, insulation and pumps. Full standby capacities are not included, pumps are in duplicate.

M   Cooling towers: forced or induced draught fans, roof-mounted cooling towers with supports.

N   *Automatic controls*: pneumatic controls including motorised valves, all thermostats, control panels, actuators, interconnecting wiring and tubing.

Ductwork

P   Supply and extract: galvanized mild steel ductwork, fittings and supports, dampers, grilles and diffusers, insulation.

Air-conditioning plant

Q   Heating and cooling batteries: humidifiers, batteries and casing and connections.

R   Fans and filters: centrifugal and axial flow fans with casings and connections, and automatic roll type filters.

S   Sound attenuation: silencers and duct lining (short lengths only).

T   *Fire detection*: heat detectors, smoke detectors, gas detectors, control panel, interconnecting wiring (excluding other fire protection services not directly associated with air-conditioning installation).

V   *Electrical work in connection*: electrical supplies to control panels and mechanical plant, mechanical services distribution board.

The number of variables which affect the cost of air conditioning is not always clearly established and the user is often unable to replace a cost based on the square metre principle in the early stages until such time as he has complete comprehensive information relating to all aspects of the proposed design. To find a solution to this problem it is necessary to break down the area cost into its various components.

The cost of the heat source will vary according to the boiler selected and the choice of fuel; the cost of ductwork will fluctuate considerably depending upon the type of system selected, the method of distribution, and the use or otherwise of the HVCA standard specification; the cost of the distribution will be affected by the introduction of pressurised ceilings and methods of control at the points of entry into air conditioned areas; the cost of cooling is affected by the design of the building fabric and choice of lighting; the cost of automatic controls depends upon the method of operation and degree of sophistication of temperature and humidity requirements; the extent of filtration and humidification can impinge substantially on the cost of air-conditioning plant and the use of the building will determine the level of sound attenuation and its concomitant effects. Each of these significant factors will require separation from the total cost if subsequent modifications and adjustments to levels of sophistication are to be taken into account with any degree of accuracy.

Each component or sub-element therefore has to be identified if its effect on the ultimate rate per m² for air conditioning is to be ascertained.

The results of such an exercise are set out in tables 11.3 and 11.4 (5). Table 11.3 relates to a variable air volume system for two office blocks of 6000 and 15,000m² respectively and table 11.4 to an induction system for the equivalent buildings. All costs are as at January 1978. The use of these tables enables each component to be reappraised immediately the design information relating to that component becomes available. The resultant effect on the cost of that component can then be realised together with its effect on the total cost and cost per m². A hitherto provisional figure for an individual component can then be replaced and the total cost per m² of the air conditioning services adjusted accordingly as a whole.

This method can of course be applied to air-conditioning installations irrespective of the area involved or type of system chosen. The systems examined to produce the data in tables 11.3 and 11.4 each proved to have various peculiarities occasioned by a restriction of the building or nature of the system required for an individual client. These peculiarities were identified and the relative cost implication removed before compiling the figures in the table and each figure therefore represents an average reflecting the design parameters outlined in the accompanying notes set out in table 11.5. In using these figures for a new project the peculiarities of any particular building or proposed design concept will require to be reintroduced and an appropriate adjustment made to the figures accordingly.

THE EFFECT OF ALTERNATIVE DESIGNS

In considering the many aspects to ensure that the air conditioning is compatible with the building many decisions are made, not the least of which is the type of system to be employed. One of the more popular systems selected five years ago related to induction systems but of late the trend appears to favour the variable air volume type of system due to the trend towards deep plan form buildings. Both these alternatives are therefore represented in tables 11.3 and 11.4

From examination of these tables it can be established that the main components affected by the choice of system are those relating to distribution and these appear to contribute most to the total elemental cost. The distribution components however often vary as much because of the planning requirements of the building as they do because of the choice of system.

For example, the ductwork distribution of a variable air volume system will normally cost three to four times the cost of ductwork distribution of an induction system. If however, a variable air volume system is installed in a simply designed building with relatively unrestricted ductwork routes and

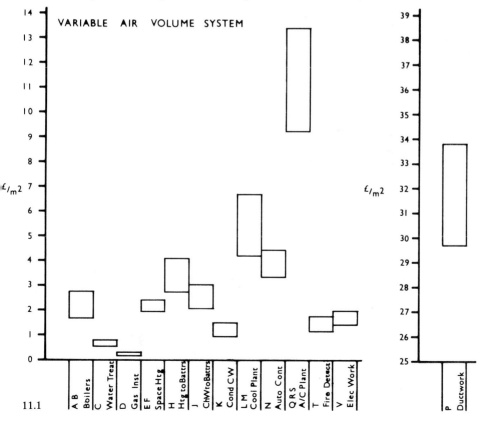

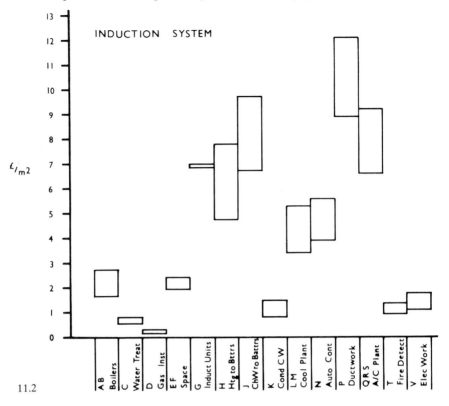

11.2

this is to be compared with an induction system installed in a building with severe planning restrictions and complex ductwork routes then the cost differential may be reduced to almost 2:1.

The isolation of the different components therefore allows those components most affected by system selection to be highlighted and a more accurate adjustment made for design decisions at the time when these are affected.

THE EFFECT OF AREA

Whilst examination of table 11.2 reveals only the indirect effect of area on the cost per m² for air conditioning, nonetheless there is a direct effect. Having divided the air conditioning element into components and using tables 11.3 and 11.4, it is possible to draw a histogram to illustrate the effect of area on each of the individual components.

Figs 11.1 and 11.2 indicate the type of result obtained. As further examples of air-conditioned buildings are analysed this type of representation could eventually be extended to show levels of complexity and sophistication of the various design system alternatives as well; as has already been established for building elements (6). The difficulty in

Table 11.6   Air-conditioning installation – variable air volume system: office block of 6 000 m²

Element	With heat recirculation		Without heat recirculation	
	Cost of element £	Cost of element per m² of floor area £	Cost of element £	Cost of element per m² of floor area £
Boilers				
Plant and instruments	11 700	1.95	12 900	2.15
Flue	4 800	0.80	5 400	0.90
Water treatment	4 800	0.80	4 800	0.80
Gas installation	1 800	0.30	2 100	0.35
Space heating				
Distribution pipework	12 000	2.00	12 000	2.00
Convectors and/or radiators	2 400	0.40	2 400	0.40
Heating to batteries	24 600	4.10	27 000	4.50
Chilled water to batteries	18 300	3.05	21 000	3.50
Condenser cooling water				
Distribution pipework	9 000	1.50	10 500	1.75
Cooling plant				
Chillers	36 000	6.00	39 600	6.60
Cooling towers	4 200	0.70	4 800	0.80
Automatic controls	26 700	4.45	26 400	4.40
Ductwork				
Supply	166 800	27.80	165 900	27.65
Extract	36 000	6.00	35 700	5.95
Air-conditioning plant				
Heating batteries	11 100	1.85	12 900	2.15
Humidifiers/cooling batteries	20 400	3.40	23 400	3.90
Fans and filters	32 100	5.35	32 400	5.40
Sound attenuation	16 800	2.80	17 100	2.85
Fire detection	10 500	1.75	10 500	1.75
Electrical work in connection	12 000	2.00	12 000	2.00
	£462 000	£77.00	£478 800	£79.80

achieving this however is due to the restricted number of air-conditioning systems available for analysis together with the differing types of building into which otherwise similar systems are installed.

THE EFFECT OF PLANTROOM LOCATION

It is often accepted that boiler plant will be situated in the basement or ground floor of a building and air-conditioning plant at roof level. The obvious advantages of siting the boiler plant near the fuel intake point and siting cooling towers and supply fans at roof level usually outweigh other considerations.

Occasionally the boiler plant can be sited on the roof with advantageous effect on heating distribution pipework and a further saving in cost if expensive excavation is not required at basement level. These savings are off-set, but only partially, by the additional cost of re-inforcing the structure to take the additional weight at high level.

Equally the chiller units and air-conditioning plant can be sited in the basement or at low level because of planning restrictions at roof level or the necessity to contain difficulties of sound attenuation to an area where the structural thickness of the walls and floors gives natural protection. This is however more expensive structurally and also large diameter pipework to cooling towers has to be extended at a proportionately higher cost.

ENERGY CONSERVATION

In selecting an air-conditioning system it is often found worthwhile to incorporate additional plant and pipework to enable heat to be re-used and thus reduce running costs and conserve energy. There are a number of methods which set out to achieve this but the type of project will generally determine at what point the increase in capital cost will be offset by the savings in running costs and the scheme still remain economically viable.

Three of these methods involve the recirculation of the extracted air in order to re-use the heat which would otherwise be expelled to atmosphere; the integration of electrical light fittings into the extract inlets which increases the amount of heat available for re-use and the reclamation of heat by the introduction of an additional heat exchanger into the pipework of the cooling water return.

The first of these methods is normally incorporated into most air-conditioning systems. If no heat were to be recirculated, the effect in capital cost terms would be as illustrated in table 11.6

The first columns of table 11.3 have been used for the purposes of comparison to highlight the effective use of heat recirculation. It is fortunately only in an extreme case that no heat is recirculated; an example of which might be the case of an air-conditioning system in an infectious diseases hospital or a case where the vitiated air contains extracted air from chemical processes or fume cupboards.

In the second method the immediate extraction of heat generated by light fittings will not only add significantly to the cooling load requirement but the additional cost of troffer units and their integration into a suspended ceiling will increase the capital cost significantly, unless offset by re-use of the heat.

The introduction of the third method will involve the installation of a water/water or water/air heat exchanger which may be of the parallel plate or tubular variety. The cost of such an innovation will be in the region of £8000 to £10,000 for a 6000m² building and generally relies upon the cooling or condenser water returning at 38/45°C to give a supply to a heat exchanger for preheating purposes of eg domestic hot water.

THE EFFECT OF HEAT RECOVERY

A further example of energy conservation is the inclusion of heat recovery plant in the air-conditioning system.

Heat recovery can be achieved with both a variable air volume system and an induction system, but is not so effective in the latter instance.

If heat recovery plant is installed in conjunction with a variable air volume system applied to a deep plan building with core areas, each probably imposing differing loads on the air-conditioning system, but all inevitably having a year round cooling requirement during occupancy; it is necessary for the central refrigeration plant to function continuously throughout the year. Therefore the heat transferred from the condenser in the refrigeration plant, which would otherwise be lost to atmosphere through the cooling tower, can be employed in winter months when heated conditioned air is required at the perimeter zones of the building to offset transmission losses through the building fabric. This can be implemented by the introduction of a double bundle condenser on the refrigeration machine used in conjunction with a heating coil sited in the perimeter zone supply air duct.

The heating coil would also be supplied with supplementary heating from the main source of heating within the building.

In the same variable air volume system, heating can also be recovered from the system exhaust air, by the use of a regenerative or recuperative heat

Table 11.7

Area of building	6 000 m²		15 000 m²	
	£	£/m²	£	£/m²
Deep plan building *VAV system* Heat pump plus heat exchanger on exhaust air	27 600	4.60	45 000	3.00
Narrow building *Induction system* Heat exchanger on exhaust air only	10 200	1.70	18 000	1.20

exchanger, which would be utilised to temper incoming fresh air for the VAV system.

When an induction system is used the induction units are generally sited at the perimeter of the building. In this location it must satisfy both heating and cooling requirements during occupancy. Where this type of system is used for air conditioning there is no continuous commitment for all year round cooling, and the only form of heat recovery that can be applied is from the system exhaust air, ie the use of a regenerative or recuperative heat exchanger as described above.

The general cost effect of introducing a basic type of heat recovery to a deep plan building with variable air volume air conditioning and to a narrow building with induction air conditioning is set in in table 11.7 (10). The office buildings used in tables 11.3 and 11.4 would require the addition of those capital costs, where heat recovery plant is to be installed.

THE EFFECT OF DOUBLE GLAZING

The cost of air-conditioning plant can be affected significantly by the extent of the cooling load. The cost of cooling an air space by 1°C can be four times as much as increasing the temperature in that same space by 1°C through heating.

The two main contributory factors which lead to the need for cooling are solar gain and heat emitted from light fittings.

In the former case solar gain can be alleviated by the introduction of double glazing, with the use of tinted, heat absorbent or reflective glass. The effect of reducing solar gain in summer and heat loss in winter will directly affect the capacity of air-conditioning plant due to the respective reduction of the maximum cooling and heating loads, resulting in a corresponding reduction in capital cost. The greater the glazed area the greater the corresponding saving will be. Similarly the greater the glazed area and the greater the sophistication in the selection of glazed units for the windows, the greater the reduction in running costs and the saving in fuel and energy requirements.

In the latter case care should be taken to avoid the tendency to provide lighting levels higher than those essential for the task required. Excess internal heat gain from lighting can substantially offset the advantage achieved by beneficial glazing combinations.

The effects of the various alternative solutions to single glazed fenestration and their corresponding effect in terms of cost are set out in table 11.8.

This table sets out alternative design solutions and gives an indication of the increase in the capital expenditure of air-conditioning plant when related to the most efficient form of fenestration. Bearing in mind that the cost of air-conditioning plant approaches a quarter of the cost of air conditioning, which in turn constitutes approximately a quarter of the cost of the complete building, the significance of the percentage increase is difficult to ignore.

Table 11.8

Ref no.	Glazing combination		Air-conditioning capital cost
1		Single sheet clear glass No shading	1.85 × £x
2		Single sheet clear glass Air space Single sheet clear glass	1.64 × £x
3		Single sheet clear glass Light colour blinds	1.52 × £x
4		Antisun grey Air space Single sheet clear glass	1.36 × £x
5		Single sheet clear glass Air space Single sheet clear glass Light colour blinds	1.33 × £x
6		Parelio polyglass grey 24 Air space Single sheet clear glass	1.19 × £x
7		Pilkingtons insulite azure Air space Single sheet clear glass	1.18 × £x
8		L H R solargrey Air space Single sheet clear glass	1.17 × £x
9		L H R clear glass Air space Single clear float glass Light colour blinds	1.16 × £x
10		Single sheet clear glass Light colour blinds in air space Single sheet clear glass	1.15 × £x
11		Antisun grey Light colour blinds in air space Single sheet clear glass	1.13 × £x
12		L H R solargrey Air space Single sheet clear glass Light colour blinds	1.10 × £x
13		Single sheet clear glass Scotchtint on inside face Air space Single sheet clear glass	1.09 × £x
14		Solarshield gold Air space Single sheet clear glass	1.08 × £x

Note: £x is the minimum cost of air-conditioning plant designed for use in conjunction with the solution ref 17.

Table 11.8—*continued*

Ref no.	Glazing combination		Air-conditioning capital cost
15		Single sheet clear glass Solarban coating on inside face Air space Single sheet clear glass	1.06 × £x
16		L H R solargrey Light colour blinds in air space Single clear float glass	1.02 × £x
17		Solarshield gold Air space Single sheet clear glass Light colour blinds	1.00 × £x

After John Bradley Associates; by courtesy of R. J. Wilson.

To establish the correct balance between capital cost of fenestration and capital cost of air conditioning plant reference should be made to table 11.9.

The cost factors a, b, c and d represent the total cost of the fenestration component named in the first column. The more complex fenestration components are expressed as a multiple of the cheaper components ie 6mm Antisun grey glass is 38% dearer than clear sheet glass.

Based on an actual example the capital cost of the air-conditioning plant was calculated for each of the 17 design solutions and can be directly related to table 11.8. Depending upon the current cost and the area involved, therefore, the relationship between the cost of fenestration and the cost of air-conditioning plant can be established for each design solution. It should be recognised, however, that the more expensive fenestration designs may not completely offset the corresponding saving on air-conditioning plant eg in one case combinations 14 and 15 proved to be the most economic solution, as combination 17 produced a higher total capital cost although in conjunction with the cheapest air conditioning plant arrangement.

Certain glazing combinations produce a mirror effect of the buildings opposite. The cost of aligning the glazing arrangement to prevent distortion has therefore to be added and this is catered for in table 11.9 by cost e.

THE EFFECT OF SHADING

There is little point in air conditioning a space if occupants are to be exposed to direct radiation from the sun in summer for long periods of time. So it is necessary to provide window shading:

on south facing facades – by horizontal overhang or set back window lines.

on east and west elevations – by vertical columns projecting beyond the window line.

Although fixed external shading is best it has not proved popular in the

Table 11.9 Schedule of air-conditioning cost and external glazing costs

Fenestration component	Glazing combination — Cost	1	2	3	4	5	6	7	8	9	10	11	12	13	14	15	16	17
Fixed aluminium external windows	a	a	a	a	a	a	a	a	a	a	a	a	a	a	a	a	a	a
6 mm clear sheet glass	c	c	c	c		c					c			c				
Opening aluminium internal window	b		b		b	b	b	b	b	b	b	b	b	b	b	b	b	b
6 mm clear sheet glass	c		c		c	c	c	c	c	c	c	c	c	c	c	c	c	c
Light colour blinds	d			d		d				d	d	d	d				d	d
6 mm antisun grey Parelio poly glass	1.38 c				1.38 c							1.38 c						
grey 24	6.20 c						6.20 c											
Pilkingtons insulite azure	5.17 c							5.17 c										
6 mm LHR solar grey	2.93 c								2.93 c								2.93 c	
6 mm LHR clear	2.72 c									2.72 c								
Scotch tint on glass	1.21 c													1.21 c				
6 mm solarshield gold	2.93 c												2.93 c		2.93 c			2.93 c
Solarban glazing units	4.91 c															4.91 c		
Air-conditioning plant capital cost [x]		1.85 x	1.64 x	1.52 x	1.36 x	1.33 x	1.19 x	1.18 x	1.17 x	1.16 x	1.15 x	1.13 x	1.10 x	1.09 x	1.08 x	1.06 x	1.02 x	1.00 x
Additional cost for alignment of reflection glazing (mirror effect)	e	e	e	e	e	e	e	e	e	e	e	e	e	e	e	e	e	e

United Kingdom because of the low solar altitudes prevailing most of the year which nullify part of its effectiveness particularly on southern exposures.

The alternative is a solution involving internal shades.

Venetian blinds have been used to effect shading with great success but this does have some effect on the external appearance of the building and louvre drapes or other drapes are sometimes preferred and each can be costed in a similar manner to the light colour blinds in the example priced out in table 11.9

THE EFFECT OF ADDITIONAL ACCOMMODATION FOR DUCT-WORK DISTRIBUTION AND PLANT ROOMS

With the introduction of air conditioning, additional area is required to accommodate the extra plant and ductwork. This additional area can be expressed as a percentage of the gross area of the building (as opposed to the treated area which is air conditioned).

The approximate percentages of the gross area required for the main types of system are set out in table 11.10. From this table the total area required for plant and ductwork appears to be in the region of 7½ to 8%. If it can be assumed that the area required for the plant and ductwork of a basic heating and extract ventilation system is in the region of 4½% then it appears that some 3% represents the additional area required for air conditioning.

The cost of this area does not attract the full average cost per m² of the total building cost. A suitable proportionate cost of a building costing £250/m² would be approximately £120/m². On this basis the 6000 m² office block would require an additional 3% of its area (or 180 m²) to be set aside for air conditioning purposes at a cost of £21,600 or £3.60 per m².

THE PROVISION OF STANDBY PLANT

It is normal for an air-conditioning system to include standby plant of some description. It is important to be able to identify the extent of this standby facility and also to consider the effect on capital cost if full standby capacity is thought to be necessary.

Table 11.10

Type of system	Area required for heating plant	Additional area for A/C plant	Area taken by terminals ducts etc
Variable air volume	2%	3.0%	2½%
Induction	2%	2.5%	3%
Dual duct	2%	3.5%	3%
Fan coil: central plant	2%	2.5%	3%
Basic heating system with extract ventilation	2%	—	2½%

Table 11.11 Provisions of standby plant
Variable air volume system with heat recirculation

Element	Cost of element without standby facilities	Cost of element in table 11.3	Cost of element with full standby capacity	Remarks on full standby capacity
	£	£	£	
Boilers				
Plant and instruments	8 300	11 700	14 050	50% increase in boiler capacity
Flue	4 450	4 800	5 200	Increase in size of individual flues
Water treatment	4 800	4 800	4 800	
Gas installation	1 400	1 800	1 850	Increase in size of fuel supplies to burners
Space heating				
Distribution pipework	11 000	12 000	12 100	Increase in size of pipework at boiler
Convectors and/or radiators	2 400	2 400	2 400	
Heating to batteries	22 600	24 600	24 800	
Chilled water to batteries	17 100	18 300	19 300	Additional pipework at extra chiller
Condenser cooling water				
Distribution pipework	8 850	9 000	9 300	
Cooling plant				
Chillers	31 000	36 000	69 000	Extra chiller required for 100% standby
Cooling towers	3 200	4 200	5 700	Upgrade cooling towers for each to handle 100% load
Automatic controls	24 900	26 700	27 000	Additional controls at chiller and boiler
Ductwork				
Supply	166 800	166 800	166 800	
Extract	36 000	36 000	36 000	
Air-conditioning plant				
Heating batteries	11 100	11 100	11 100	
Humidifiers/cooling batteries	20 400	20 400	20 400	
Fans and filters	28 400	32 100	32 100	
Sound attenuation	16 800	16 800	16 800	
Fire detection	10 500	10 500	10 500	
Electrical work in connection	10 900	12 000	12 100	Increase in cable sizes
	£440 900	£462 000	£501 300	
Cost per m²	£73.48	£77.00	£83.55	

Again using the 6000m² office block containing the variable air volume system it is possible to isolate the standby facility on the one hand and then to compare the result if full standby capacity is provided on the grounds that a deep plan building can be rendered uninhabitable in the event of a complete breakdown of the primary plant.

Table 11.11 compares the cost of the elements firstly where no standby facility is provided and secondly when full standby capacity is provided.

PROVISION FOR ALTERNATIVE FUELS

The recent energy crisis has resulted in a re-examination of the possibility of introducing boiler plant capable of utilising more than one fuel.

The two most common fuels considered in this respect are oil and gas because of their present economic viability. The introduction of dual fired boilers is not so much to take advantage of beneficial fuel tariffs as to build in a standby facility in the event of failure of the primary fuel source.

The effect on capital cost of introducing a dual fuel facility is set out in table 11.12. This table again relates the cost to the air-conditioning systems for the 6000m² office block in tables 11.3 and 11.4.

The main increase in capital expenditure can be determined from a comparison of the first or second columns with the third column, where a more significant difference occurs when the boiler plant is increased to accommodate the dual fired burner. The minor differences elsewhere are self explanatory.

Table 11.12   Capital costs for alternative fuels

	Gas fired		Oil fired		Dual fired	
	£	£/m²	£	£/m²	£	£/m²
Boilers						
plant & instruments	11 700	1.95	14 090	2.35	15 360	2.56
Gas installation	1 800	0.30	—	—	1 800	0.30
Oil installation (including oil tanks)	—	—	1 550	0.26	1 550	0.26
Automatic controls	26 700	4.45	26 800	4.47	26 800	4.47
Electrical work	12 000	2.00	12 300	2.05	12 300	2.05

RUNNING COSTS

The four main contributory factors to be taken into account when assessing running costs are the cost of electricity, heating fuel, maintenance cost and depreciation of capital.

An earlier article by R. C. Cracknell (7) considered the cost factors in running air conditioning in five government offices in Central London and came to the conclusion that the annual cost of energy varied between £2.52 and £4.83 per m² (January 1978 costs).

Table 11.13 Comparative cost indices for comfort air-conditioning systems

System	Capital cost index	Electrical running cost index
VAV: medium velocity	115	100
Chilled ceiling plus low-velocity air	128 (1)	122
Versatemp plus air through wall locally	103 (3)	133
Perimeter induction: 2-pipe, non-changeover	100 (2)	155
Fan coil (4 pipe) plus low-velocity air	107 (3)	218
Dual duct: high velocity	132	378

Notes: (1) Includes the cost of the ceiling.
      (2) Excludes the cost of unit casings.
      (3) Excludes the cost of electric wiring.

As with capital costs there are many factors which influence running costs. These include the area of the building, the proportion of glazed area, the choice of single and double glazing, shading, type of air-conditioning system, choice of automatic controls, the provision of standby facilities, type of heat source and the orientation and disposition of the building. In fact, most of the factors which affect capital cost also impinge upon running costs. It is therefore sometimes more difficult to forecast and assess running costs when in addition the figures obtained from historical data also have to be rationalised against a background of actuarial calculation and fiscal expediency.

The relationship between different systems in terms of capital cost will not be the same when the comparison is made for running costs. Based on a recent study (8) table 11.13 gives an indication of the ratios of capital cost and electrical running costs for comfort application in a typical office building.

To give an indication of running costs on a comparative basis in terms of cost related to area, reference can be made to table 11.14. The first column indicates the cost per m² for the cost of electricity consumed on an annual basis. The second column indicates the cost of heating fuel, and has been standardised to highlight the effect of the variation in choice of system. The third column gives the total cost of energy consumed for the building. The cost of water consumption and gas for catering is generally sufficiently insignificant that it may be discounted when considering air-conditioning running costs.

Table 11.14 Comparative running costs for air-conditioning systems

	Electricity used		Heat used		Total cost
	kWh/m²	p/m²	kWh/m²	p/m²	p/m²
Single duct VAV with perimeter heating	50	135	115	85	220
Single duct VAV with terminal re-heat	75	205	85	65	270
Dual duct VAV	75	205	85	65	270
Dual duct constant volume	150	410	115	85	495

It is assumed that a gas fired boiler provides the heat. Electricity is assumed to cost 2.5p/kWh as a flat rate and natural gas about 15p/therm, representing about 0.5p/kWh which, for a mean seasonal boiler efficiency of 70%, gives about 0.7p/kWh.

These figures are essentially the results of an examination over a range of figures where extreme examples have been discounted and an average taken of the remainder.

The examination of a particular example, a twelve storey office block containing approximately 14,000m² and incorporating a two pipe induction air-conditioning system produced the undernoted figures.

The total energy consumed was estimated at 1.47 million kWh and cost £36,750 annually for electricity which reflects a cost of 2.5p/kWh or £2.625 per m² at January 1978.

This electrical consumption included refrigeration, compressors, pumps, all fans, lighting and business machines. Lighting consumption could be based on 20 watts/m² and an allowance for business machines on approximately 5 watts/m². The lighting costs appeared to require 50% of the total electricity cost or approximately £1.3 per m², leaving the cost of electricity for the air-conditioning system at £1.325 per m².

The cost of heating fuel, which was based on gas at 17p/therm, came to £17,066 annually or £1.22 per m².

Maintenance costs amounted to approximately £5 per m², which is equivalent to almost twice the cost of electricity consumption for the air conditioning system and the cost of heating fuel together.

If one were to allow depreciation of capital at a discount rate of 10% on the basis of a 20 year life to complete the calculations for running costs then an amount of £8/£9 per m² could be allowed.

Total running costs for this building would then amount to £16 per m².

Operating costs were surveyed in some detail for some 25 commercial office buildings and presented in a paper produced in 1971 (11). Here the relationship between the various components was recognised and also the effect of window area and air conditioned floor area/gross floor area. The buildings in this paper defined air conditioning as heating, ventilation and cooling by mechanical methods to provide comfort conditions for the staff. A similar exercise to reflect present modern air-conditioning standards would therefore be extremely worthwhile.

THE RELATIONSHIP BETWEEN CAPITAL AND RUNNING COSTS

It is important to be able to relate an increase in capital expenditure to a saving in running costs. In some cases the saving in running costs will not justify the proposed increase in capital expenditure unless certain economic factors are taken into account. In such cases these factors may appear to have an indirect influence only on what is otherwise a basic decision on services design. The establishment of the relationship between capital and running costs depends upon capital cost calculated at the present time being compared with running costs to be expended in the future calculated in equivalent present day terms.

In order to evaluate running costs in terms of the above, account has to be taken of the present value of £1 for running costs forecast for succeeding years, the cost of funding capital expenditure at appropriate interest rates, techniques for assessing comparative costs such as discounted cash flow, fiscal policy regarding allowances on plant and machinery, levels of corporation and other taxes and the life of individual items of plant and the residual value at the end of their useful life.

A paper produced in 1976 by P. H. Day (12) sets out in specific terms the applications of discounted cash flow techniques to engineering services plant and highlights the effect of fiscal legislation. The commercial decisions to be considered by senior management when taking into account risk, uncertainty and inflation are also listed, including balancing the rise in fuel and energy costs against the possible fall in the value of money. In this respect it is worthwhile contemplating the possibility of legislation being introduced to conserve energy resources in a similar manner to the way building regulations and planning approval requirements operate at present.

A paper produced in 1977 by D. A. Butler (13) explained the investment appraisal technique of cost benefit analysis. This sets out with the objective of achieving comparability and to provide an assessment of economic viability expressed in terms of a common base. Costs are therefore presented utilising the technique for discounted cash flow, together with a sensitivity analysis to highlight the effect of changes in the estimated values which would otherwise reflect perfect market conditions. In this latter respect it should be recognised that the discounted cash flow technique in itself does not take into account the effects of inflation (14) or the strength of sterling in world markets.

The application of these techniques does provide a detailed appraisal to substantiate the increase in capital expenditure because of future savings in running costs, or demonstrates that such expenditure would be unjustified.

An initial appraisal however often has to be made on fairly approximate figures and a refined technique is sometimes too delicate an instrument in such circumstances.

In broad terms therefore if the capital expenditure on the air-conditioning plant for one of the systems described earlier in tables 11.3 or 11.4 were to be increased by £10,000, the running costs would require to be reduced by at least an average £900 per annum (or £0.15/m² per annum) before such a decision became economically viable.

The above calculations assume certain predetermined conditions. Capital is raised on the basis of a mortgage and requires to be repaid over 20 years at a rate of interest of 10%. Repayments are alleviated by relief on corporation tax which it is assumed would otherwise be levied in full at the rate of 52%. Air-conditioning plant is regarded for tax purposes as fixtures and fittings rather than as plant and machinery and has an assumed life of 20 years. Running costs are calculated on the basis of a 10% rate of inflation per annum.

CONCLUSION

In a paper produced in 1964 (15) by C. P. Swain, D. L. Thornley and R. Wensley, which reviewed air-conditioning systems current at that time, a statement was made that a comparison between systems is only really valid if it compares 'like with like' and that a choice of systems is subject to economic, planning and many other pressures. It was also stated that clients and others were not always familiar with the provisions necessary to accommodate air conditioning which had resulted in some unnecessarily high costs up to that time.

Unfortunately the implications of these statements are still as valid today, mainly due to the increasing difficulty of comparing air conditioning systems on a 'like to like' basis.

In order to control the costs of air conditioning all the factors which prevent direct comparison have to be identified, isolated and understood. Then the techniques described in this chapter which enable the costs to be monitored in a practical manner can be applied.

In applying these techniques at the appropriate stages of design development, installation and operation, the designer has the opportunity to appreciate the cost effect of his decisions. Effective cost control can then be realised and where the demand for energy conservation is apparent the necessary calculations to achieve the optimum economic solution will be made in sufficient time for the correct result to be achieved.

REFERENCES

1   P. E. Bathurst and D. A. Butler, *Building Cost Control Techniques and Economics,* London, Heinemann, 1973, p 132-140.

2   A. E. Turner and P. C. Venning, Cost Control of M & E Services, *Chartered Surveyor: Building and Q.S. Quarterly,* vol 3, no 1, 1975, Appendix 4.

3 H. G. Mitchell and J. Leary, Environmental Quality and its Cost, *Integrated Environment in Building Design,* (ed) A. F. C. Sherratt, Barking, Applied Science Publishers, 1974, pp 250-273.

4 W. P. Jones, Designing Air-Conditioned Buildings to Minimise Energy Use, *Integrated Environment in Building Design,* (ed) A. F. C. Sherratt, Barking, Applied Science Publishers, 1974, pp 171-206.

5 Davis, Belfield and Everest (ed), *Spon's Mechanical and Electrical Services Price Book 1978,* London, Spon, 1977, pp 201-204.

6 Davis, Belfield and Everest, Initial Cost Estimating: The Cost of Warehouses, *The Architects' Journal,* 1 June 1977, pp 1037-1042.

7 R. C. Cracknell, Cost Factors in Owning and Running Air Conditioning, *Integrated Environment in Building Design,* (ed), A. F. C. Sherratt, Barking, Applied Science Publishers, 1974, pp 218-229.

8 By courtesy of W. P. Jones and Haden Young Ltd.

9 J. D. M. Robertson, Cost and Quality Data Banks in Operation, *Quality and Total Cost in Buildings and Services Design,* (ed) D. J. Croome and A. F. C. Sherratt, Lancaster, Construction Press, 1977, pp 143-151.

10 By courtesy of R. E. J. Shave and Building Design Partnership.

11 N. O. Milbank, J. P. Dowdall and A. Slater, Investigation of Maintenance and Energy Costs for Services in Office Buildings, *Building Research Establishment Current Paper, CP38/71.* December 71, JIHVE, Vol 39, October 1971, pp 145-154.

12 P. H. Day, Capital Investment Appraisal for Mechanical and Electrical Services in Commercial Buildings, *The Heating and Ventilating Engineer,* February, 1976.

13 D. A. Butler, Ambient Energy – Economic Appraisal, *Ambient Energy and Building Design,* (ed), J. E. Randell, Lancaster, Construction Press, 1978, pp 29-38.

14 B. Carsberg and A. Hope, *Business Investment Decisions under Inflation,* The Institute of Chartered Accountants in England and Wales, 1976.

15 C. P. Swain, D. L. Thornley and R. Wensley, The Choice of Air-Conditioning Systems, *J. Inst. Heating Ventilating Engineers,* vol. 32, April, 1964, p 307.

DISCUSSION

H. Thomas (Donald Smith, Seymour & Rooley)   I would like to refer to two factors which should be considered very carefully during initial design and cost forecasting. The two factors are system commissioning and future plant maintenance. We are concerned with energy conservation and in these two areas I feel a considerable amount of energy is continuing to be lost. Are the present standards of system commissioning and plant maintenance adequate?

R. B. Watson (Davis, Belfield & Everest)   This raises a point of particular interest. When assembling the costs for this paper I was unable to identify from the historical data at my disposal any of the maintenance or commissioning costs separately. In most cases when contractors are tendering they include the costs for commissioning and testing in their other prices.

It is therefore very difficult to isolate these particular costs and all the costs in the paper include for the cost of any commissioning or testing etc. It has nonetheless become increasingly obvious to me, particularly in the case of competitive tendering, that a contractor tends to underestimate the cost of commissioning and perhaps this aspect of the question can be referred to both Mr Anthony and Mr Ginsler for further comment.

E. J. Anthony (W. S. Atkins & Partners)   There is certainly need for more attention to be paid to what is required at the commissioning stage, firstly by really good definition of the functional operation of the systems, which is not always done to a satisfactory degree in tender documents. A commissioning document should be produced by the designer which defines all the sequences of commissioning required to prove that the plant performs the duties which were originally required of it in the design brief.

There can be very great difficulties particularly with large jobs which take a long time in that the conditions change from the initial conceptual stage to the commissioning stage. In the case described in my paper we were particularly unfortunate in the period spanned by our commissioning programme. I do not think it is unique to find that buildings do not achieve their predicted loads. Building complexes with a number of buildings, in particular, do not achieve the loads predicted and there are difficulties in commissioning the plant and attempting to operate it, certainly in the early years, under conditions which are never those at the design stage. Much attention must be paid to careful sifting of information. Computer plant is an area where loads are seldom if ever achieved. Once a plant is installed for a particular load a great deal of redesign is necessary to reoptimise for economy in energy use, requiring variation in plant sizing, motors, etc. No contractor or consultant wants to redesign a job without a fee half way through the commissioning stage because the functions for which the building was originally defined as being required no longer exist.

L. S. Ginsler (Rybka Smith & Ginsler Ltd)   All of us are very much aware of the problem of maintenance and operating costs. We have tried to specify wherever possible devices which are self-balancing, using factory pre-fabricated units as much as possible to reduce to some extent the commissioning problem. We also have introduced in our specifications a requirement for any of the problem devices – such as an emergency generator plant – to be tested in the manufacturer's plant in our presence before delivery to the site. It is so much easier to carry out the test and any corrections needed while it is still in the factory than after it is on the site. Since we have instituted that method of specification there have been fantastic improvements in the construction programme and an incredible reduction in the number of problems. Some of the problems encountered before instituting this method were almost beyond belief, for example, a compressor fitted with the wrong impeller, which was jammed inside the casing.

As part of the energy conservation programme we instruct staff in the preparation of a graphic wall chart to indicate monthly energy consumption. This gives a very simple dramatic visual aid to identify possible consumption problems.

Maintenance is always a problem. We have tried everything, including outside service organisations and a full time maintenance staff for the building. We have never really achieved an entirely successful programme.

Recently control equipment manufacturers have gone into the business of maintenance and they are very successful. They have the knowledge of how the system is supposed to work, they have an interest, the intelligence and the training. So far we are delighted with the result because it has taken a tremendous load off our shoulders – whether it will degenerate to the level of past performance when the initial enthusiasm wears off we must wait and see. An important part of the maintenance problem is the 'mechanical nightmare' of our own creation. We should not forget what I call the 'KISS' principle – 'Keep It Simple Stupid'.

Dr D. Fitzgerald (University of Leeds)   I would like to compliment Mr Watson. The data he has given us is what we have been waiting for. It is the kind of information needed at the beginning of a project, which hitherto has not been available.

12 Predicting annual energy consumption

M. J. HOLMES

INTRODUCTION

The rise in fuel prices over the last few years has increased interest in the energy consumption of buildings. One of the major energy consuming items in many buildings is the heating and air-conditioning system (1). Not all systems are perhaps as efficient, in energy terms, as might be desired and there is therefore potential for energy savings. The first step in determining any possible energy saving is the estimation of the energy requirements of a system. In general the designer will be faced with two situations:

- a proposed building where details of plant and system are not clearly defined
- a proposed system where plant and control schedules are known.

A different approach may be required for each case.

Broadly speaking there are three possible approaches to the calculations of the energy consumed by the HVAC system:

(1) Determine the equivalent full load hours of operation of each plant item. (Equivalent full load hours is equal to the number of hours that plant item would have to operate at full load to consume the same amount of energy when operated under typical conditions for a complete year).

(2) Determine 'system efficiencies'. That is use factors to represent the overall efficiency of a complete system when operated under typical conditions for a complete year.

(3) Construct a detailed mathematical model of the system and use a computer to calculate the performance of the model when operated under typical conditions.

The first method has great attraction because of its obvious simplicity, and is part of the method given in IHVE *Guide Book B* for estimating the energy consumption of heated and air-conditioned buildings. Its drawback is the inherent simplicity, it is unlikely that such a method can be used if it is required to demonstrate the finer points of different systems.

The second method, would allow distinction between various system types but pre-supposes that calculations have been carried out for a wide range of systems. This data is not yet available.

The third possibility is the most flexible, and has very few restrictions. It does of course assume access to a computer and can be expensive to use.

The possibilities of each method are discussed here, with the most emphasis on the use of computer simulation. This is because of the obvious advantage of the ability to study, in detail, the differences between systems. It is clearly unsuitable for use at the conceptual stage of the building where method 1 has obvious advantages. This chapter, which is based upon BSRIA research work (2) discusses some aspects of each method and in particular outlines problems associated with the three approaches.

EQUIVALENT FULL LOAD HOURS

The demand on an air-conditioning system can be expected to depend upon the building in which it is installed, thus the following are important:
- glazing, area and type
- orientation
- shape
- internal gains – influenced by the building usage and illuminance
- outside air requirements
- infiltration

IHVE *Guide Book B*, section B18, gives a single figure of 1040 equivalent full load hours for refrigeration plant for all types of systems. Is this really acceptable when all types of buildings and plants are to be covered? Data presented by the Electricity Council (3), based on studies or existing buildings and systems shows a spread, in equivalent full load hours, for refrigeration plant from about 750 to 1500 hrs. The 'conventional' 1040 hrs would therefore give errors in the region of -40% to $+30\%$, which could be considered excessive. Refrigeration energy is however, only one part of the total building usage, IHVE *Guide Book B* indicates that it is only about 10% of the total electrical energy demand. In this case errors on overall consumption would reduce to the region of $\pm4\%$ which is small. Thus it may well be reasonable to use the equivalent full load hours concept when considering the overall energy demand of a building. If the system is to be considered as a separate entity the larger errors may well be unacceptable. Any closer estimate will require an assessment of individual systems.

SYSTEM EFFICIENCY METHOD

This approach can be considered similar to the equivalent full load hours approach as it attempts to use a blanket parameter to cover the performance of a system. The difference between the two methods lies in the word system which must pre-suppose a difference between systems.

It is unlikely that a constant efficiency can be assigned to a system, thus the system efficiency will be a function of the load imposed by the building. The problem is to obtain values of efficiency for various systems and load patterns. This is demonstrated in fig 12.1 where each point represents the calculated thermal load on the cooling coil for a particular building thermal load for:

(1) A dual duct system with a wild cooling coil and a fixed outside air quantity of 20%.

(2) The same basic dual duct system with the outside air quantity controlled to minimise the refrigeration load.

There is a fairly well defined relationship between building load and refrigeration load for the poorly controlled system but this is not so when optimisation is attempted. These results, and others, which were obtained from simulation studies, suggest that the system efficiency approach may not be viable. If the method is to be used then a large number of systems should be examined either using measurements on actual buildings or computer simulations.

COMPUTER SIMULATION METHODS

This method has virtually no restrictions providing the characteristics of each system component are known. It is however more complex than the other two possible approaches, and therefore it is worthwhile to examine both the method and factors that may influence the results.

What is simulation?
Simulation in its simplest form is the representation of any physical system

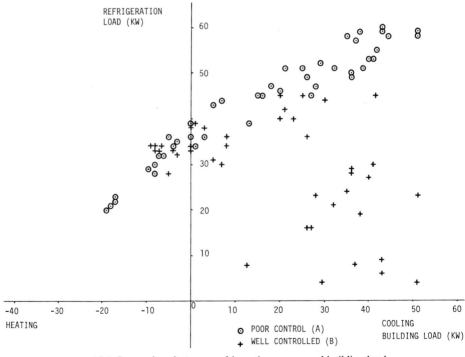

12.1 Comparison between refrigeration power and building load

by a model. If an input is applied to the model then the output from the model will, if the model is sufficient for the required purpose, represent how the system will behave in practice. The output can only be satisfactory if both the input and the model are representative of expected operating conditions.

The important elements of simulation are:

- specification of output required; this might be temperature, heat load or energy consumption
- specification of input, that is the parameters which are expected to affect the output, perhaps weather, lighting loads etc
- specification of a model suitable for the required objective. This model will normally comprise a set of equations.

Simple example of component simulation

A system simulation will normally comprise a set of component simulations, the output from one component forming the input to the next element in the system. The validity of the total simulation will depend upon the quality of the component models. These models should be of sufficient detail to adequately represent the component behaviour.

Consider a variable speed fan linked to a variable flow rate supply system, an assessment of the annual energy consumption is required. The inputs to the model are:

- air flow distribution pattern, ie the number of hours at particular flow rates.
- the system pressure loss as a function of flow rate.

The above define the required energy, the model must represent the conversion between desired performance and input energy necessary to achieve this. The model must also take account of the control system applied to the fan.

The output will be the energy input to the fan.

We will consider several methods of assessing the energy consumption.

(a) Take no account of variable flow. The maximum design flow rate is 10m³/s at 1500 Pa, and the fan runs for 5110 hrs per year. These are the inputs. The model requires fan and drive efficiencies. These can be obtained from catalogue data. The fan efficiency is 75% and the motor drive assembly is 80% efficient, the annual energy consumption is then obtained from the model:

$$\text{Energy} = \frac{\text{Pressure rise} \times \text{flow} \times \text{time} \times 3.6 \times 10^{-6}}{\text{fan efficiency} \times \text{motor \& drive efficiency}} = 456 \text{ GJ} \quad (1)$$

(b) The energy calculated under (a) is clearly an overestimate because the input was not that to be expected under practical conditions, but the model was satisfactory for the required purpose. A realistic input would be the temporal distribution of air flow rate, this can be obtained from the load pattern imposed by the building structure, weather and usage

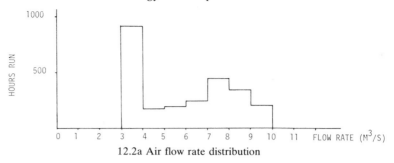

12.2a Air flow rate distribution

(this will be discussed in a later section). Figure 12.2a shows the distribution for this example. The system and control will be assumed perfect,
that is system pressure loss is proportional to flow rate squared, and
the fan speed proportional to flow rate. Thus the fan efficiency is constant.
The motor and drive efficiency will be assumed proportional to speed
which is reasonably representative of some drives and motors.

   The energy is calculated using Equation 1 for the range of flow rates,
and running times given in fig 12.2a. The resultant annual consumption
is 188 GJ. This is very much less than that of the constant flow system.
(c) The result obtained under (b) is still not a very good estimate of the fan
   energy consumption. Whilst the input was similar to practice the model
   was not. In practice a controller would be used to maintain a constant
   pressure at a reference point in the system. As resistance is increased
   by the terminal units throttling the flow rate, the controller will reduce
   the fan speed to return the control pressure to the set point. Taking the
   set point as 500 Pa, the system pressure-flow relationship will be that
   shown in fig 12.2b.

   A new model is now required. For each flow rate the corresponding pressure loss can be obtained from fig 12.2b, the fan characteristic
will enable the efficiency and speed to be determined as a function of

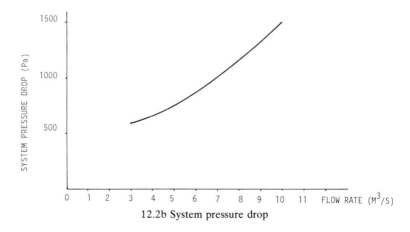

12.2b System pressure drop

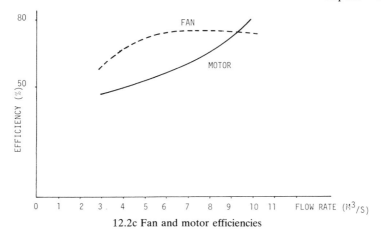

12.2c Fan and motor efficiencies

flow rate. The motor and drive efficiency can also be specified as a function of a flow rate. An efficiency model can be drawn up as in fig 12.2c, and the input power calculated for each flow rate using Equation 1. The resulting annual energy consumption is 238 GJ, 27% higher than that with perfect speed control.

(d) The model represented by method (c) is probably sufficiently close to practical conditions to make further sophistication unnecessary. The next stage in the elaboration of the model would be to include the effect of a proportional band on the controller. A new pressure flow relationship will be required and, then without going into details, if the proportional band is 100 Pa then the predicted annual energy consumption will increase to 258 GJ.

The example has demonstrated the effect of different models on the output, the completeness of the final model will depend upon the desired accuracy, and the cost of the simulation exercise.

INPUTS

The concept of input – model – output is the basis for all simulation work. This paper is mainly concerned with plant and system simulation, the inputs are therefore:
- the thermal loading
- external temperature

The thermal load, which may be both sensible and latent, will depend upon the building type, usage and weather conditions. External temperature, which effects the outside air loading on the system, is solely weather dependent. It is therefore pertinent to discuss the choice of representative weather conditions, building usage and the most suitable methods for calculating the building loads. As it is not the object of this paper to present a detailed analysis of the methods for calculating the building loads, only the general requirements will be indicated here.

Weather data
Most system designers are familiar with the weather data contained in the IHVE *Guide*, this data, with perhaps the exception of degree day information is not suitable for the calculation of annual energy consumption, or studies of the behaviour of a system. Design data, by its very nature, is representative of extremes. Fig 12.3 gives an example of the results obtained when room loads are calculated using:
• March cooling design data
• March heating design data
• an average March day

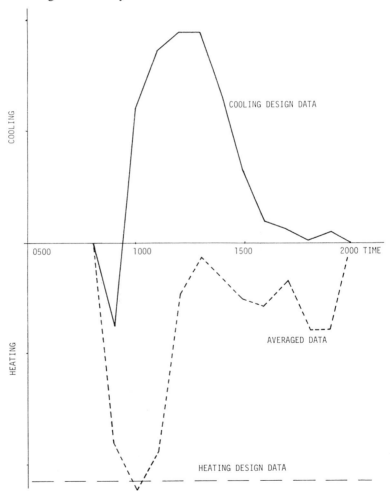

12.3 Plant loads in March

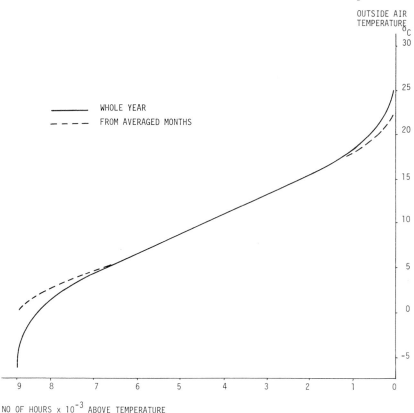

OUTSIDE AIR
TEMPERATURE
°C

WHOLE YEAR

FROM AVERAGED MONTHS

NO OF HOURS x 10^{-3} ABOVE TEMPERATURE

12.4 Yearly distribution of outside temperature

Whilst none of the loadings is likely to give a true representation of the required delivery in March, the average day will probably be closest to the truth. The load patterns shown in fig 12.3, indicate the problem in air-conditioning simulation. Averaged data can give a bias towards one particular mode of operation, and produce results that are not representative of the normal usage of system ie in the present example only heating is used in March. The only satisfactory solution is to use loads calculated for every hour of the year, this is probably not practicable. Errors introduced by averaged data are indicated by the ogives given in figs 12.4 and 12.5. BSRIA research work (4) has indicated that heating energy can be estimated from averaged months (that is taking one month's real weather data and averaging down to produce an average day).

A similar exercise has not been carried out on air-conditioning systems because the increased complexity will make a general conclusion impossible. It is probably best at present to either accept the biased results of averaged monthly data or to use a complete year. An alternative to these is to calculate the year's thermal loading and then to treat each month separately using a statistical load pattern, ie so many hours at full load, so many at 90% etc. External temperature variations during the month do not make this a simple exercise.

If all simulations were carried out with the same weather information, then it would be feasible to analyse this data to produce statistically acceptable months. A second, and probably the most important advantage of using identical weather data throughout the industry is that all simulations would be related to the same base. A method for selection of representative weather data is given in (5).

Internal gains
Weather is an external input or loading on the building and system, internal gains are a further input which cannot be ignored. If the energy estimate is to be realistic then a typical usage of the building should be assumed. It is therefore necessary to specify hourly levels of occupancy, lighting and the

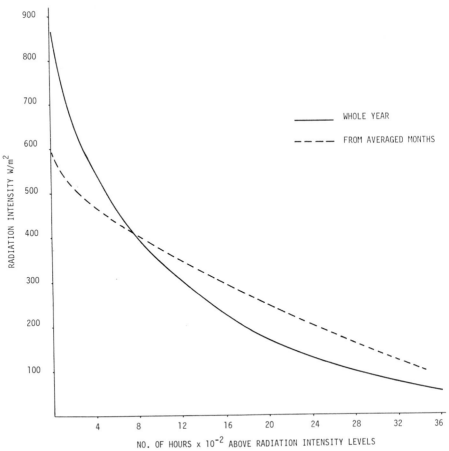

12.5 Yearly distribution of intensity of total radiation on horizontal surface usage of internal equipment. Of these, lighting offers most difficulty.

With the exception of the core of a deep plan building, it is unreasonable to assume that the lights are on for all the occupied period. IES *Technical Report No. 4* (6) suggests that internal lighting will be used when the illuminance available from daylight falls below that provided by the artificial lighting. This has been confirmed by a recent survey (7). It can therefore be assumed that if sky illuminance, the reflective properties of the internal surfaces, and the transmission characteristics of the windows are known, then daylight factors (8) can be used to predict the internal illuminance levels. If these levels are below the design lighting level then the lights can be assumed to be on.

The only practical difficulty involved in estimating the availability of daylighting is the sky illuminance. Whilst standard values can be used for lighting design purposes (6), the type of simulation considered here requires typical levels dependent upon the weather. It is suggested (8) that the luminous efficiency of natural daylight is about 120 lumens/W, thus the sun/sky illumination can be calculated from solar radiation data. Figure 12.6 shows the monthly distribution of lighting energy for an office of 15m² floor area with 30% single glazing, internal illuminance of 400 lux, and lighting rated at 20 W/m². The diagram compares the consumption for continuous operation of the lights, lighting switch using solar design data derived from IHVE *Guide Book A* and switching dependent on averaged monthly real weather data.

Plant and system input – building thermal load

The building thermal load, together with external air temperatures are the normal plant/system inputs. The plant output will be either energy or some desired room or system conditions. If energy is the output then room temperature becomes an input. The installed plant size can be expected to have some influence on room temperature, especially when the plant is

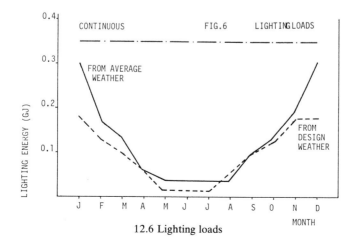

12.6 Lighting loads

slightly undersize. Even with an adequate plant size, the response time and controls may affect building's temperatures, and through this the building thermal load.

General requirements
The simplest possible model of a building is that used to size heating plant, ie heating load = UAΔt. + infiltration loss (or gain)*. This type of model is only suitable for steady state calculations, that is when internal and external conditions remain constant. Hour by hour calculations cannot be done with such a simple model because heat is stored in the building fabric. Crude estimates of annual heating system consumption can, however, be carried out using the steady state model, as described in IHVE *Guide Book B* (1). It has already been suggested that this method gives no insight into the behaviour of the plant or the effect of different types of control. If a reasonable representation of the real situation is to be studied then the model must allow calculations at a sufficient number of intervals to produce a realistic load distribution. The result will then allow the system performance calculation to take due account of plant efficiency at loads other than design. It is usually accepted that the smallest practical interval is one hour.

Basic consideration
The first essential is that the loads on separately controlled zones are isolated. The required system delivery to each zone, is required, not the total building load; in the extreme the total building loading could be zero with the system supplying equal heating and cooling to two zones. Whilst it is not the object to recommend a particular programme or calculation method, it is pertinent to discuss the general approach to programme selection.

The building simulation suggested in the IHVE *Guide* (which is for design loads) is based on the pseudo steady state response to a sinusoidal loading. That is external conditions are assumed to be adequately represented by a sine wave with a 24hr period, and to repeat for a time interval that is long compared with the building response time. The internal temperature should be constant (24hr plant operation). Given these conditions the calculated loading will be fairly representative of practice. Real weather does not follow a sine wave. Further, most buildings do not employ 24hr plant usage, the room emperature will, therefore, not be maintained at comfort levels over a 24hr period. Thus a programme based on constant room temperature and steady sinusoidal weather is, in general, unsuitable for typical system loading calculations for use in the calculation of energy consumption.

It may also be necessary to consider the interactions between plant operation, control, and building loads. Intermittent operation of a heating system with a fixed start time will result in a different load pattern from that arising from the use of an optimum start system (4). The type of room

* U = Thermal transmittance W/m²°C, A, surface area m², Δt inside air to outside air temperature difference °C

control may also influence the energy required from the system. Results from a simple model of system and building with different room controls are shown in table 12.1.

Table 12.1   Effect of room control on energy requirements

Building response time	Annual energy consumption (GJ) control type		
(Notional)	Perfect	Proportional	On/off
Long	7.2	8.2	7.3
Medium	6.8	8.0	6.7
Short	6.4	6.3	6.3

These results are for an intermittently heated building (17hrs daily operation). Perfect control means that a constant room temperature of 22°C is held throughout the occupied period. Proportional control has a set point of 23.5°C with a proportional band of 3°C (to give an average room temperature of about 23.5°C). The on/off controller had the same range. The main reason for the differences shown in table 12.1 is that different room control systems result in different room temperatures, thus the building load is not the same in each case. Similar effects are found with heating plants having different response times. An example is given in table 12.2.

Table 12.2   Effect of plant response time with proportional control

Plant response time (hrs to full output)	Annual energy consumption (GJ) building type Heavy	Medium	Light
0	7.9	7.5	5.9
0.2	8.2	7.8	6.1
0.8	8.2	7.9	6.3
1.6	8.2	8.0	6.3
3.2	8.2	8.0	6.3

The differences in consumption with varying plant response time shown here do not appear significant.

These results are for a long period (14–16hrs) of occupancy. If the occupancy time is shortened to about 8hrs then fig 12.7 is more representative of the effect of plant time constant. It follows that the control system and plant time constant both effect the required energy and the load programme should take account of this. Most systems will have a time constant of about ½ hr (system response time to full output of 2hr).

It may be concluded that the building thermal load calculation should involve plant response characteristics, plant control and typical operating

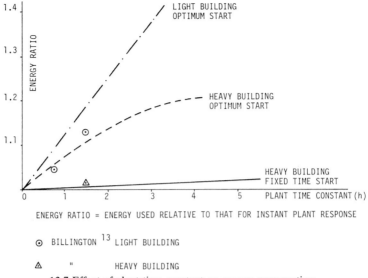

ENERGY RATIO = ENERGY USED RELATIVE TO THAT FOR INSTANT PLANT RESPONSE

⊙ BILLINGTON [13] LIGHT BUILDING

△   "   HEAVY BUILDING

12.7 Effect of plant time constant on energy consumption

conditions. Figure 12.8 shows the type of result which may be obtained from such a programme.

So far no mention has been made of either system type or infiltration. The type of system employed will only affect the thermal loading if it causes a change in room conditions. This means that we only need to consider three basic types:

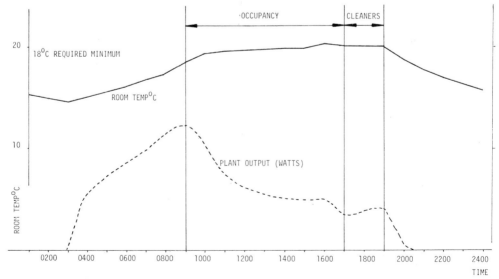

12.8 Room temperature and plant output as calculated from combined building and plant simulation programs

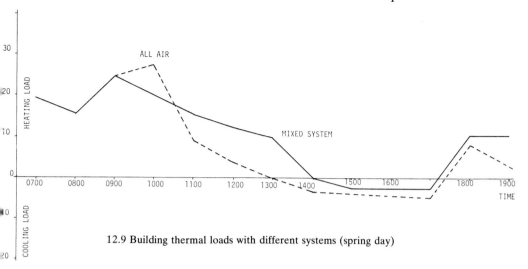

12.9 Building thermal loads with different systems (spring day)

- wet system   heating via convectors or radiators only
- all air system   room loads balanced by air supplied at a temperature or flow rate controlled by a room thermostat
- air water system   room loads split between two systems, ie fabric losses taken account of by skirting heating on an external temperature compensator, air supplied for cooling

An example of cooling loads calculated for a spring day with an all air and air/water system are shown in fig 12.9. The two systems maintain different space temperatures and consequently the system demand requirements are not the same.

Infiltration is not easily calculated. A rather complex and , from the view of hourly load calculations, expensive programme is required (10). Simple assumptions such as infiltration rate proportional to wind speed (obtained from weather data) may cause (in the author's opinion) errors as great as those from an assumption of a constant infiltration rate. This is because of the influence of stack effect, variable openings and the probable unknown pressure distribution on the building. Thus although infiltration should be taken into account, great sophistication at excessive operating costs may not be worthwhile.

Summing up, the important features of a building thermal load programme for use as input to a system simulation are:

(1) Thermal storage must be modelled.
(2) The input must be suitable for real as opposed to design weather data.
(3) Plant shut down, overnight and weekend, must be taken into account.
(4) The maximum plant size must be an available input.
(5) The plant response should not be assumed instantaneous.

(6) It should be possible to take some account of the overall control of the plant.

(7) It would be an advantage if the fundamental differences between systems could be taken into account, ie whether the output of one part of the system is solely dependent upon outside conditions.

(8) Infiltration must be allowed for when expected in the real building.

COMPLETE AIR-CONDITIONING SYSTEMS

The large number of plant items and available systems (arrangements of plant items) result in fairly complex simulation routines. The modelling of individual items is fully discussed in (2). The major difficulty associated with the calculation of the duty of an air-conditioning plant is caused by latent loads. If these could be ignored then the problem would be simplified. Thus one must ask the question 'is an accurate determination of the latent load necessary for the estimation of overall energy consumption?' The answer, with the exception of process air-conditioning, is probably 'no'. If the refrigeration energy is equal to the heating energy then assuming the latent load is zero, when it really comprises 30% of the cooling load, will result in an overall error of 15%, this may not be acceptable. A more reasonable assessment in typical air-conditioning situations would be that the latent load is no more than 20% of total refrigeration load. Assuming a sensible heat ratio of 0.9 overall will only introduce errors in the region of 5% which is probably significantly less than the error in the overall energy estimate. It is therefore probable that latent heat transfer can be ignored in comparative energy consumption calculations. Such calculations will be termed 'simple simulations'. Simple simulations should only be used if:

- latent gains are not large compared with overall sensible loading
- outside air quantity is *not* controlled via an enthalpy controller
- no evaluation of the quality of room humidity is required
- no account is taken of controller proportional band.

We will first discuss the type of work that can be done with complex simulation, and then give some examples of the use of simple simulation. The main advantage of simple simulation is that it is very cheap, and therefore many more system possibilities can be assessed.

Complex simulation

A complex simulation will take account of all relevant factors in the system, from duct heat losses through to a detailed calculation of the cooling coil latent loading.

Energy consumption of a two zone air-conditioned office

This example, which should not be generalised, considers the air conditioning energy consumption for a two zone office with 24hr plant operation. Complex simulation is used to examine various plant and control options.

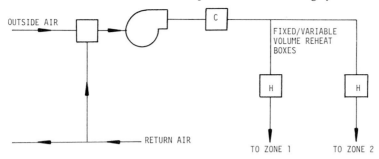

12.10a Outline of single duct system

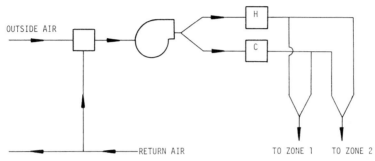

12.10b Outline of dual duct system

The energy consumption figures are based on loads representing a typical day from each season of the year.

In all cases the system was to maintain an internal dry bulb of 22°C with a fixed air quantity of 0.5m³/s. The fan was sized at 5.8m³/s and 1250 Pa static pressure. Thus for a fan static efficiency of 75% and a motor/drive efficiency of 78% the power consumption for 96 hours running (at full flow) is 1191 (GJ).

Brief description of test runs
The following systems were investigated (see fig 12.10).
F1 Single duct reheat. Perfect control of zone reheater, wild cooling coil.
F2 Single duct reheat. Perfect control of zone reheaters. The off cooling coil dry bulb controlled at 15°C.
F3 Dual duct. Proportional controller on heater. Set point 33°C with a 5°C proportional band. Wild cooling coil.
F4 Dual duct. As F3 but with perfect control of cooling coil dry bulb at 15°C.
F5 Variable flow rate. System flow ratio 1.5:1. Zone reheaters have perfect control and only function at minimum flow rate. Wild cooling coil.
F6 Variable flow rate. As F5 but with perfect control of cooling coil dry bulb at 15°C.
F7 Variable flow rate. As F5 but flow ratio 2.3:1.

F8   Variable flow rate. As F6 but flow ratio 2.3:1.
F9   Variable flow rate. As F8 but with boiler capacity reduced by 50%.
F10 Dual duct. Perfect control on heater and cooler with set points of
     33°C and 15°C respectively.
F11 Dual duct. Perfect control on heater and cooler with set point of 28°C
     and 15°C respectively.

Calculated energy consumption

Table 12.3 shows the system energy consumption as calculated by the
simulation programme.

Table 12.3

Run no	Boiler GJ	Refrigeration plant GJ	Fan GJ
F1	18.3	4.5	4.3
F2	17.8	4.5	4.3
F3	12.0	2.9	4.3
F4	9.9	2.5	4.3
F5	15.8	3.0	1.3
F6	11.1	2.1	2.4
F7	11.6	2.4	0.9
F8	7.2	1.8	1.7
F9	6.8	1.8	1.7
F10	10.0	2.6	4.3
F11	7.5	2.2	4.3

Note These figures do not include the energy consumption of associated
        pumps and the cooling tower.
Whilst runs F1–F11 only cover a small number of possible system and
control options for the building a few relevant points may be made.

With reference to table 12.3:

(i) *Use of wild cooling coils*
The saving of capital cost due to the omission of controls in the cooling coil
may be justifiable under some circumstances.
    The single duct systems F1 and F2 show only 3% and 2% increase in
boiler and refrigeration plant consumption respectively when a wild coil is
used. (This is probably because these systems are very inefficient.) A more
dramatic increase is apparent with dual duct systems F3 and F4 (30% boiler,
16% refrigeration plant), and the variable flow rate systems F5 and F6 (42%
boiler, 43% refrigeration plant). The latter increases suggest that wild coils
are likely to be uneconomic, however examination of the full results shows
that the effective set point for the wild coil was about 11°C, thus indicating
an oversized coil. More accurate sizing would reduce the difference between
controlled and uncontrolled coils.
    The choice of cooling coil set point is somewhat dependent upon the

degree of control required over relative humidity. The test results did not show unacceptably high humidity levels with a 15°C set point.

(ii) *Use of improved controls*
The proportional controller employed on the heating coil had a proportional band of 5°C. This means that the set point will only be achieved when no heating is required, and at full output the off coil air temperature will be 5°C below this set point. Thus with a design coil off temperature of 28°C and a proportional band of 5°C the set point is effectively 33°C. Comparison of dual duct runs F4, F10, F11 where proportional control and perfect control at upper and lower limits of the proportional band are simulated, shows that closer control can result in significant energy savings.

Simple simulation
It has already been suggested that errors associated with the assumption of a constant cooling coil sensible heat ratio may not be significant. Some of the systems described in 4.1.2 were simulated using the simplified technique, ie only considering sensible heat balances and perfect control. The differences between the two, with a constant sensible heat ratio of 0.9, are shown in table 12.4.

Table 12.4 Comparison of simple and complex simulation

Run no	Boiler input (%)	Refrigeration input (%)
F2	−0.6	−8.4
F11	+1.7	+15.8
F9	+2.6	+10.5
F6	+0.8	−1.4

The % difference between simulations was calculated as:

$$\left(\frac{\text{Energy used in simple simulation}}{\text{Energy used in complex simulation}} - 1 \right) \times 100\%$$

Fan consumptions were the same for all cases.

Whilst individual differences may be considered large the maximum effect on overall energy consumption, if refrigeration and fan energy is five times more expensive than boiler fuel, is less than 10%. Such comparisons can only be applied to systems with perfect control.

The following example demonstrates the use of simple simulation to establish the best control strategy for the plant.

Comparison of control methods for a simple variable flow rate system
An office building comprising two zones is to be air conditioned using a variable flow rate system. The system has skirting heating controlled on an external dry bulb compensator (set for zero heat output when the outside

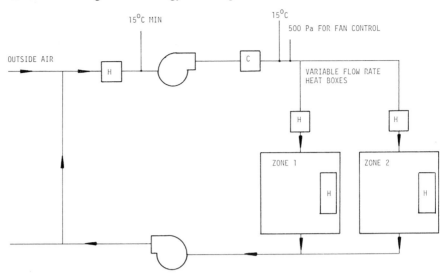

15°C MIN

15°C
500 Pa FOR FAN CONTROL

OUTSIDE AIR

H

C

VARIABLE FLOW RATE
HEAT BOXES

H

H

ZONE 1

ZONE 2

H

H

12.11 Outline of variable flow rate system

temperature is 14°C) with cooling from the system shown in fig 12.11. The design conditions are:
- fan – 19.5m³/s at 1250 Pa static pressure
- maximum flow: minimum flow, 2.5:1
- outside air quantity 4m³/s
- cooling coil set point 15°C

A variable speed fan drive is to be used with motor/drive efficiency proportional to speed. Control is via a pressure sensor set to maintain 500Pa.

The following system possibilities are suggested.

SV1 no additional control

SV2 a heat recovery system between exhaust and intake air ducts. Efficiency of 0.7, and a pressure loss of 200 Pa at 4m³/s

SV3 control of outside air quantity on the following schedule

dry bulb	outside air quantity
0-14°C	minimum
14-21°C	no recirculation
>21°C	minimum

SV4 install monitoring control system to isolate the cooling coil if both zone supply temperatures are greater than that measured upstream of the cooling coil.

The annual system energy consumption is shown in table 12.5 (loads are based on hourly average for each month of a typical year). Note that the overall coefficient of performance of the refrigeration plant was 3.

Table 12.5   System energy consumption (GJ)

System	Boiler	Refrigeration plant	Fan
SV1	1592	156	328
SV2	1490	335	446
SV3	1597	145	328
SV4	1420	113	328

The example shows that:

(a) An efficient system is not greatly improved by additional control.

(b) Heat recovery, as used here does not necessarily save energy. The use of the compensated heating system minimises any advantages of heat recovery. (See also note below.)

(c) The increased system resistance due to the heat recovery device, increases fan and cooling energy.

(d) If the basic system were less efficient, then some of the control options would show significant savings. Assuming a simple single duct reheat system (ie VAV with max to min flow ratio of 1.0) then the annual energy consumption for such a system, with the same control options SV1–SV4 is given in table 12.6.

Table 12.6   Annual energy consumption (GJ)

System	Boiler	Refrigeration plant	Fan
ISV1	3250	765	864
ISV2	3250	988	1021
ISV3	3250	577	864
ISV4	2097	422	864

The results shown here are for example only, they should not be generalised.

Note The results displayed in tables 12.5 and 12.6 indicate a large increase in refrigeration energy when the heat recovery system is employed. This is because there is no control on the system to restrict its operation when cooling is required at low outside air temperatures. The increase in fan energy and hence fan temperature rise also contributes towards the increase in refrigeration energy.

CONCLUSIONS

Three possible methods have been presented that may be used to predict the energy consumption of air-conditioned systems. It is suggested that only detailed simulation using computer techniques can show the differences between systems. It is however possible to make an approximate estimate of the overall energy consumption of an air-conditioned building by means of the concept of equivalent full load hours of plant operation.

REFERENCES

1   IHVE *Guide Book B,* section B18, 1970.
2   M. J. Holmes, *BSRIA Project Report 15/111,* 1978 (restricted circulation).
3   Electricity Council (Environmental Engineering), *Energy Consumption of Air-Conditioned Buildings,* October, 1973.
4   S. Adams, Effect of Plant Oversize and Operating Schedule on the Energy Consumption of a Heating System, *BSRIA Project Report 15/115,* 1977 (restricted circulation).
5   M. J. Holmes and E. R. Hitchin, An 'Example Year' for the Calculation of Energy Demand in Buildings, *Building Services Engineer,* Vol 45, No 10, January, 1978, pp 186–189.
6   Daytime Lighting in Buildings, *IES Technical Report No 4,* 1972.
7   L. J. Stewart, Private communication.
8   R. G. Hopkinson (et al), *Daylighting,* London, Heinemann, 1966.
9   H. J. Butler, Approximation of the Electrical Load of Lighting Installations, *BSRIA Technical Note 6/76,* 1976.
10 I. N. Potter, *Operating Manual for the CRKFLO Computer Programme Report No 15, 1934/3* (confidential to DHSS), August, 1976.

BIBLIOGRAPHY

The references given here contain useful information concerning simulation procedures and results. They are in author alphabetical order.

J. S. Askari, Universal building complex load and energy simulation programme, *Heat Pip Air Condit,* September, 1967, **39** (9), pp 106–112.

Beckett & Hart, *Numerical Calculations and Algorithms,* McGraw-Hill, 1967.

D. M. Black, A study of the dynamic behaviour of a boiler plant using hybrid computing techniques. *J Inst Fuel,* November, 1970, pp 43, 358, 467–475.

D. Bridges, Computer programmes point the way to energy conservation, *Heat Pip Air Condit,* January, 1974, **46** (1), pp 93–97.

S. Y. S. Chen, Existing load and energy programmes, *Heat Pip Air Condit,* December, 1975, **47** (13), pp 35–39.

W. J. Coad, The computer as a tool for energy analysis, *Heat Pip Air Condit,* January, 1975, **47** (1), pp 46–50.

C. P. Crall, Procedure for the simulation of the refrigeration system at the Ohio State University test site, paper presented at ASHRAE semi-annual meeting, 23–27 January, 1975, New Orleans.

W. C. Dries, Heat gain and heat loss calculation by computer, *ASHRAE J,* March 1966, **8** (3), pp 74–77.

W. E. Eavers, Ecube computerised energy analysis: energy–equipment–economics, *ASHRAE J,* September, 1971, **13** (9), pp 46–53.

K. N. Feinderg, The use of computer programmes to evaluate energy consumptions of large office buildings, *ASHRAE J*, January, 1974, **16** (1), pp 73–76.

N. E. Hager, Computer method for estimating net energy requirement for heating building, paper 24, symposium on use of computers related to buildings, National Bureau of Standards, Gaithersburg, November/December, 1970.

E. Hallanger, Further comments on the Kansas City computer simulator programme, *ASHRAE J*, October, 1974, **16**, (10), pp 58–60.

D. E. Hamilton (et al), Dynamic response characteristics of a discharge air temperature control system at near full and part heating load, *ASHRAE* (trans), 1974, **80**, part 1, pp 181–189.

J. D. Haseltine and E. B. Avale, Comparison of power and efficiency of constant speed compressors using three different capacity control methods, *ASHRAE* (trans), 1971, **77**, part 1, pp 158–162.

M. J. Holmes, Part load efficiency of gas and oil fired boilers, *BSRIA Technical Note 1/76*, 1976.

U. Inoe and H. Lee, Simulation of refrigeration system for energy conservation and the results verified by actual measurements (trans), *SHASE*, No 1, 1976, pp 93–106.

R. W. James and S. A. Marshall, Dynamic analysis of a refrigeration system, paper presented to Institute of Refrigeration, 18 October, 1973.

C. D. Jones (et al), Simulation and verification of an hydronic sub-system with radiant panels. *ASHRAE* (trans), 1975, **81**, part 1, pp 457–462.

C. D. Jones (et al), Simulation and verification of a direct expansion refrigeration system, *ASHRAE* (trans), 1975, **81**, part 1, pp 475–483.

C. D. Jones (et al), Summary results of the Ohio State University field validation programme, *ASHRAE* (trans), 1976, **82**, part 1, 340–346.

C. D. Jones, A computer simulation and validation of a building heating, ventilating and air-conditioning system, *ASHRAE* (trans), 1975, **81**, part 1, 506–518.

W. Jones (et al), Simulation of a dual duct, reheat, air-handling system, *ASHRAE* (trans), 1975, **81**, part 1, 457–462.

J. J. Kowalezeski, A method of predicting the performance of unit air conditioners at different outdoor and indoor conditions, *Aust Refrig Air Condit*, January, 1969.

P. Kruger, Computer simulation and actual results of an energy conservation programme, *ASHRAE J*, October, 1974, **16** (10), pp 52–57.

M. Lokmanhekim and R. H. Henninger, Computerised energy requirement analysis and heating/cooling load calculations of buildings, *ASHRAE J*, April, 1972, **14**, 4, 25–333.

A. C. Pittas and R. T. B. McKenzie, Digital computer analysis of the performance of a refrigerant forced draft cooler, *ASHRAE* (trans), 1971, **77**, part 2, pp 180–187.

E. B. Qvale, The development of a mathematical model for the study of rotary-vane compressors, *ASHRAE* (trans), 1971, **77**, part 1, pp 225–231.

W. F. Stocker, *Procedures for simulating the performance of components and systems for energy calculations,* third edition, *ASHRAE,* 1975.

ACKNOWLEDGEMENT

This work is reported by permission of the Building Services Research and Information Association who gave permission to use the figures.

DISCUSSION

R. C. Legg (Polytechnic of the South Bank).   To make the economic comparison between systems, we need to be able to calculate the annual energy consumption and its cost. However, it is often sufficient to compare only the differences between systems and not the total cost of the complete system as for example with heat recovery from the exhaust air where there is a preheat requirement in the supply air. In this case we compare the additional capital costs with the savings in annual energy costs. For this a method is required which lies somewhere between the methods outlined in Mr Holmes's chapter. Equivalent full load running hours is too insensitive for accurate analysis, whereas a computer might be a very sophisticated sledgehammer.

The method I advocate was originally described by Robertson* and

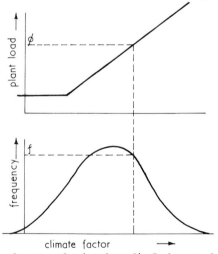

12.12 Plant load, Φ, and frequency of occurrence, f, plotted against climate factor (R. C. Legg)

subsequently developed†. It is a technique junior engineers can readily apply (I teach it to diploma students at the Polytechnic). The method is to establish the relationship between energy demand and an appropriate climate factor and then to integrate this with the annual frequency distribution of the climate factor. This is illustrated in fig 12.12 where plant

* P. Robertson, Evaluation of air conditioning energy costs, *Building Services Engineer,* 42: 195, 1974.
† R. C. Legg, Analysis of energy demands of air conditioning systems, *Heating and Ventilating Engineer,* vol 51, July, 1977, p 6.

load Φ is plotted against the climate factor for which there is a known percentage frequency of occurrence, f. The average hourly energy demand E. is then given by the numerical integration

$$E = {}^{1}/_{100} \Sigma (f \times \Phi)$$

From this calculation the annual energy costs may be obtained e.g. for heating using fossil fuel

$$\text{Annual heating cost} = \frac{E \times h \times F a\ 100}{\eta \times CV}$$

and for a refrigeration plant

$$\text{Annual refrigeration cost} = \frac{E \times h \times G}{COP}$$

Where h   =   hours of operation
F   =   fuel cost per unit mass
G   =   electricity cost per kwh
CV   =   calorific value per unit mass
COP   =   coefficient of performance
η =   boiler firing efficiency

Until recently appropriate data on climate factors was not readily available to enable the engineer to proceed with an analysis of energy demands. However, this has now been partly rectified with the publication ‡ of frequency distribution tables of the hourly values of outside air dry bulb, wet bulb and enthalpy. These tables are based on a series of reports § published by the Meteorological Office for 13 stations in the UK. A typical distribution based on ten years observations, is reproduced in table 12.7 for

Table 12.7   **Frequency 'f' of annual hourly values of outside air dry-bulb Temperature, t_0-Croydon, UK (R. C. Legg)**

t_0 °C	f %	t_0 °C	f %	t_0 °C	f %	t_0 °C	f %	t_0 °C	f %	t_0 °C	f %
−20	−	−10	0.01	0	2.21	10	6.13	20	1.80	30	0.05
−19	−	−9	0.02	1	2.77	11	5.99	21	1.39	31	0.02
−18	−	−8	0.03	2	3.23	12	5.74	22	1.07	32	0.01
−17	−	−7	0.05	3	3.49	13	5.60	23	0.77		
−16	−	−6	0.08	4	3.82	14	5.21	24	0.53		
−15	−	−5	0.18	5	4.33	15	4.96	25	0.36		
−14	−	−4	0.35	6	4.92	16	4.37	26	0.28		
−13	0.01	−3	0.66	7	5.26	17	3.73	27	0.20		
−12	0.01	−2	0.98	8	5.97	18	3.08	28	0.12		
−11	0.01	−1	1.58	9	6.16	19	2.39	29	0.07		

‡ R. C. Legg and P. Robertson, The frequency of occurrence of hourly values of outside air conditions in the United Kingdom, IEST. Polytechnic of the South Bank, *Technical Memorandum No 22,* 1976; M. Holmes and S. Adams, Coincidence of dry and wet bulb temperatures, BSRIA *Technical Note 7N 2/77,* 1977.
§ Meteorological office, *Combined distribution of hourly values of dry bulb and wet bulb temperatures,* A series of reports covering 13 stations in the UK, HMSO.

dry bulb temperature at Croydon. Because these figures include measurements for every hour of the day, they are applicable strictly only to continuously operated plants, but they may also be applied to intermittently operated plants provided a good proportion of the load can be related to outside air conditions. More extensive climatic data from 23 stations in the UK is now being analysed for publication in the near future.

Dr L. Longworth (University of Manchester, Institute of Science and Technology)   Mr Holmes made a very convincing case for using a computer programme to assess annual energy requirement. He said the method had virtually no restrictions provided the characteristics of each system component were known. Is there sufficent data available about the partial load characteristics of components and is research under way to produce more?

M. J. Holmes (Ove Arup Partnership)   There is some information available on the part load characteristics of plant items. Boilers can be represented by a plotting part-load efficiency against load. Building Services Research and Information Association have published a technical note on this subject (Technical Note 1/76), which not only compares available data but also gives the part load characteristics for a typical boiler. The Building Research Establishment have also published similar data. There are some very good models of refrigeration plant available. The problem with refrigeration plant is that a detailed model could be more complex than one for a complete air-conditioning system and there is evidence to suggest that an overall mean coefficient of performance may provide an acceptable model. The problem is discussed in some of the references given in the paper.

Manufacturers' catalogues are useful for the characteristics of items such as fans, pumps, heat exchangers etc. The data can be curve fitted to put it in a form suitable for simulation programmes. There are of course some problems in assessing the electrical input to the fan motor as the installed motor capacity may differ significantly from the fan requirements. I think there may be difficulties with some of the heat recovery devices as catalogue data is often based on conditions that differ from those normally found in the UK.

G. Mole (Bernard Sunley & Sons Ltd)   You seemed to indicate that even a well controlled heat regenerator has negative benefit. I believe that used in the appropriate application they are beneficial.

M. J. Holmes   I did not mean that heat recovery devices are not worthwhile. I was attempting to demonstrate that you cannot just add a device to a system and expect to save energy without considering how it is controlled.

The example chosen was a very efficient air-conditioning system, and once you have high efficiency it is very difficult to improve. Of course these devices save energy, but it is necessary to make a study of how it will function in the complete system. It is also worth noting that if the efficiency of a heat recovery device is 70%, 70% recirculation will achieve a similar energy saving without the penalty of increased fan power.

E. C. Lovelock (Shell UK Administrative Services)   When buildings are occupied for short periods, say eight or ten hours per day, five days per week, what is the best method of dealing with the loss of heat during the shut-down period whilst the building is not occupied? Does Mr Holmes' computer model present the economics of installing additional plant capacity to give boost heating for a short period of time prior to occupation on the basis of intermittent heating as against the older system of 'night setback'?

M. J. Holmes   An answer will be published very shortly by BSRIA*. Part of this whole project was in fact an examination of intermittent heating using simulation methods and in particular to study the optimum start controller, but I did not present that work here because it would take as long as this paper. Obviously there are advantages in intermittent heating and this type of simulation programme can be used to determine the energy savings.

P. R. Fish (Atkins Research and Development)   How in practice would you use a computer on a real building bearing in mind that each floor may consist of perhaps a dozen zones, and every floor may be different, making the whole problem extremely large and complex. There is also the problem of how many different months to examine in the year.

M. J. Holmes   I am coming round to the view that it may not always be practicable to use the computer in the detail we would like. I think that some of the routines involved in the complex simulation programme may result in excessive computation costs, especially if the building model includes a representation of every room. The only satisfactory method would appear to be to use experience to reduce the number of sub-divisions to an acceptable level. I cannot see any other answer at present. It is obviously not possible to just take the whole building shell as under some circumstances zero energy would be predicted when in practice equal cooling and heating are required.

I think that it is necessary to have a method of representing each month in the year. Whether this is by calculating loads for every hour of the month is another question. Studies, at BSRIA, of heating systems suggest that an average day will yield satisfactory results. This is probably because the energy consumption of a heating system is very closely related to the external dry bulb temperature. The energy consumption of conditioning systems depends upon wet bulb, dry bulb and solar radiation, and averaging a month down to one day may cause significant errors. In this case it may be necessary to examine the operation of the system at every hour of the year. If this is thought to be impractable, then at least the results obtained from some averaged real weather data will be better than any calculated from design data.

M. Corcoran (Building Design Partnership)   Mr Holmes described the difficulties of using computer techniques for the thorough analysis of the energy pattern of use of systems in buildings and expressed reservations about using the estimated full load running hours as an approach for

* S. Adams and M. J. Holmes, *Effect of Plant Oversize and Operating Schedule on the Energy Consumption of a Heating System,* BSRIA Project Report 15/115, 1977.

selecting between alternatives. Does this not suggest that for the time being at least we must concern ourselves with thorough energy analysis of buildings – using one of the number of computer programmes available and with the operating parameters of systems? We should not attempt to establish codes related to overall building energy use which may develop into mandatory codes.

M. J. Holmes   The disadvantage of simple methods for predicting energy consumption is that they cannot take full account of all features of the system. Simulation studies similar to those described in the paper have shown that it is possible to distinguish, at a very simple level, between various air-conditioning systems. Because refrigeration energy is only a small proportion of the total building energy, whether it is worthwhile distinguishing between systems for the purposes of energy target calculations can be argued.

J. W. Coe (Standard & Pochin)   Mr Holmes' example of heat recovery on the simple variable flow rate system leads more to the need to associate heat recovery with good controls than to the conclusion that the heat recovery system indicated is inefficient. A heat recovery device added to a system will impose an additional resistance on the fan which will reduce its volume and also energy consumed. I feel sure Mr Holmes would not want to cast doubt on the usefulness of various proved heat recovery components which can be very useful under carefully considered circumstances. In particular in winter there is heat input from the air which would reduce the required boiler capacity.

M. J. Holmes   I must emphasise that the examples given in the paper are only intended to demonstrate the use of simulation and should not be generalised. If a heat recovery device is added without increasing the fan performance air flow will drop, and there will not be such a large increase in fan energy as shown in the paper. The reduction in air flow rate would mean that the two systems (one with heat recovery, one without) are not necessarily comparable because, for example, the room air movement would change. If internal conditions are not changed, then there will be a significant increase in fan energy. Also, in this particular example, refrigeration load is increased because of an increase in the temperature rise across the fan.

The examples given in the paper cover a complete year. When there are no controls on the heat recovery device, if there is high solar gain and a low air temperature, heat will be transferred from the extract air to the supply air even if cooling is required. This will increase the load on the refrigeration plant.

J. W. Coe   In a simple system the refrigeration plant could be switched off in the winter. There must then be some benefit from the heat recovery.

There are now various types of heat recovery devices available and I am concerned that the detailed reference to a duct mounted device, if taken out of context would appear to be discouraging.

M. J. Holmes   Yes, of course, but whilst it would be possible to give a large number of examples covering every eventuality, I have taken a few examples and indicated what might happen in specific cases. To some extent I am saying 'Look you can do this and sometimes the result may surprise you'.

The results of further work on the run-around coil system will be available in the near future* which shows calculated energy savings and penalties due to increased fan power and pump power.

*M. J. Holmes and G. Hamilton, *Energy Savings with Run-around Coil Heat Recovery Systems,* BSRIA Technical Note, 1978.

13 Review of systems against energy criteria

J. P. G. GOLDFINGER

INTRODUCTION

The energy consumption of a building is one of many factors that must be considered by the designers of buildings. In air-conditioned buildings the air conditioning itself is a major energy consuming factor. It is not possible to review all the combinations and permutations of air-conditioning systems for a variety of building types. The systems reviewed are those which have been used in the UK for office buildings.

The type of building for which all the systems could be used is typically a shallow plan building of, say, anything from 3 to 12 storeys, with a modular configuration with the facility of individual control of the temperature conditions for each module or group of modules. The approach used can be applied to other air-conditioned buildings and other air-conditioning systems.

As in a business operation it is possible to identify profit centres, in an air-conditioning system it is possible to identify energy centres. Profit centres/energy centres are identifiable but they do not operate in isolation from each other. The analogy can be continued by saying that profit or energy can be transferred from one centre to another.

In the air-conditioning systems the energy centres are:

- energy source for the building
- energy conversion plant
- energy distribution systems
- the terminal device.

These can be represented graphically as in fig 13.1.

GENERAL REVIEW OF ENERGY CENTRES

Energy source for the building

The selection of the energy source for any building is based on availability, convenience and cost. The two basic sources for air-conditioning systems are usually fossil fuels such as oil or gas for heating and electricity for motive power and sometimes heating. Other sources such as high temperature hot water or steam are sometimes available for groups of buildings or districts.

Whereas fossil fuels are delivered to site and the efficiency of their use is solely dependant on the energy conversion within the building, electricity at

250

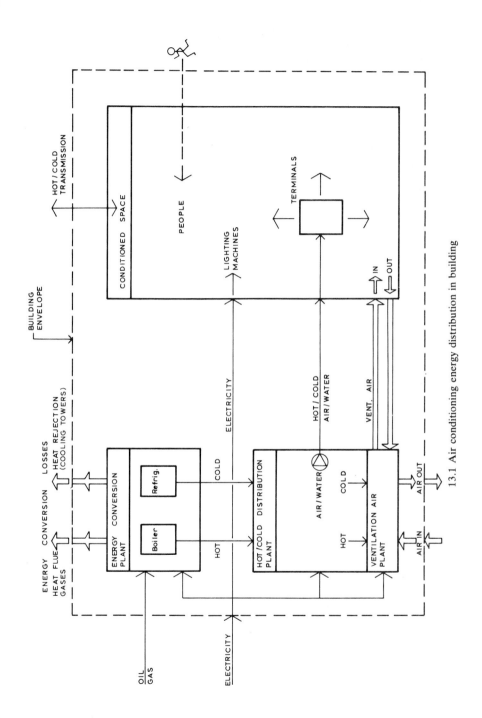

13.1 Air conditioning energy distribution in building

the building boundary is the result of the conversion of fossil fuels or nuclear energy and its distribution. The efficiency of this is about 27%.

Energy conversion plant

For air-conditioning systems energy is converted into heated and chilled media. In all the systems reviewed except one this is done in a central plant, the exception being the reverse cycle system. Heat is generated by boiler plant with an average seasonal efficiency of 70% and cold is generated by vapour compression refrigeration water chillers with an average coefficient of performance of 3.5 to 1. Again the exception to this is the reverse cycle system which operates at about 2.5 to 1. Heat rejection from the building is through cooling towers.

For maximum energy conservation in this energy centre, the conversion plant must be matched carefully to the air-conditioning system it serves, to ensure that it operates at maximum efficiency. Any system which dissipates the converted energy is obviously wasteful.

There are ways in which energy that is normally thrown away from the central plant can be recouped. Heat that is rejected through cooling towers can be diverted to heat conditioning air. This possibility can rarely be justified on cost grounds, however, because of the small amount of energy recovered.

The amount of air being handled by heat rejection plant justifies detailed consideration of the location of this plant to see if it cannot be used for ventilation of spaces such as garages.

Energy distribution

From the energy conversion plant the hot and cold water has to be distributed throughout the building to the terminal equipment. At the same time air must be distributed for ventilation. The method of distribution together with the terminal units define the air-conditioning system.

The medium used for this distribution is either air or water, except in the case of the reverse cycle system. In this latter system most of the energy distribution is electrical as the energy conversion to hot and cold takes place at the terminal itself.

As a generalisation it is possible to transmit heat and cold with less energy by pumping water rather than by pumping air because of the higher specific heat of water. The variable air volume system (VAV) does, however, overcome this particular disadvantage of air distribution systems.

The aim of the energy distribution and ventilation systems must be to ensure that the hot and cold generated by the central energy conversion plant is transferred to the conditioned space as effectively as possible. This means:

(a) That in the distribution plant itself consideration must be given to ensuring that there is no unnecessary heating, cooling and reheating of air. Careful attention must be given to the psychometrics of a system to ensure this, together with a realistic approach to the ultimate conditions of temperature and humidity to be achieved in the conditioned spaces.

(b) That the losses between the distribution plant and the terminals are kept to a minimum. These losses are

• the friction losses of the medium in pipes or ducts
• heat losses from the pipes or ducts, which are kept to a minimum by keeping the temperature difference between the medium and the spaces through which it passes as low as possible and by good insulation
• actual leakage of the air from the ductwork.

(c) That in the case of air systems there is no unnecessary waste by exhausting air rather than recirculating it, or where possible heat recovery systems are used to retain the energy within the building envelope.

The terminal device

This is the element of the whole system which is closest to the ultimate user. The more control he has of it the more potential there is to save energy by switching it off when he does not need it.

The terminal device can vary from a complete refrigeration unit as in the case of the reverse cycle system to an air outlet with a damper automatically controlling the amount of air supplied to the conditioned space.

It is the terminal unit itself that finally dictates the efficiency of the system. If air that has been cooled is then reheated at the terminal there is obvious waste. If a terminal is able to transfer energy from the space to the system that can be used elsewhere in the building the system is energy conserving. If the terminal is able to deliver just the right amount of heating or cooling to the space without wasteful losses within itself, the system is energy conserving.

AIR CONDITIONING SYSTEMS

The seven air-conditioning systems reviewed have advantages and disadvantages beyond those of pure energy considerations. Some take up less space, some are cheaper, some easier to install and commission, some easier to maintain, and so on.

All the systems give
• adequate ventilation
• adequate cooling
• adequate heating
• possibility of control of temperature of each module
• good air distribution within the space at acceptable noise levels.

The seven systems reviewed are
• reverse cycle unitary system
• four pipe fan coil system

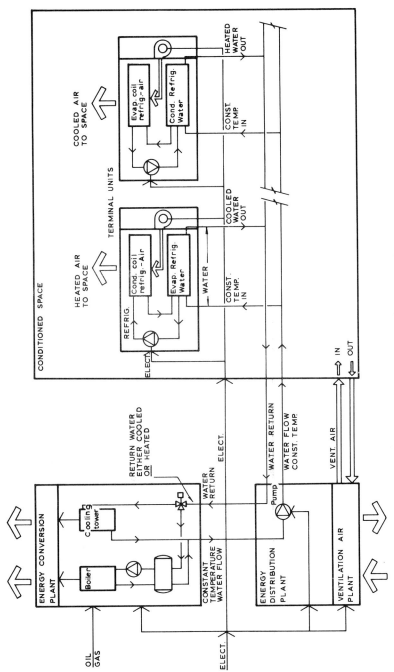

13.2 Schematic presentation of reverse cycle system

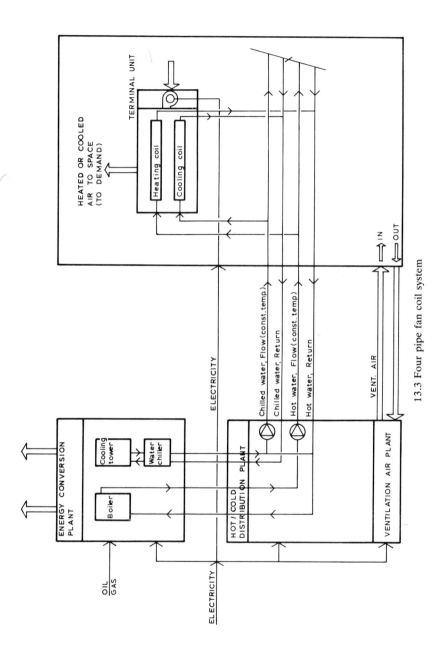

13.3 Four pipe fan coil system

Table 13.1   Approximate proportion of annual energy consumption for air-conditioning systems

Oil or gas for heating	30% to 50%
Electricity for refrigeration fans and pumps	70% to 50%

Note: This does not include electricity for lighting

Table 13.2   Typical proportions of electrical consumption of two pipe induction system, non-changeover

Central water chilling plant		*55%*
Water chiller	45%	
Cooling tower	5%	
Condenser pumps	5%	
Distribution system		*45%*
Fans	35%	
Pumps	10%	

Tables abstracted from *Energy consumption of air-conditioned buildings.* Electricity Council 1973.

- two pipe non change-over induction system
- four pipe induction system
- central station all-air system with terminal reheats
- dual duct system
- variable air volume system.

In reviewing the systems it is worthwhile bearing in mind the proportion of energy used in each energy centre. This is given typically in tables 13.1 and 13.2. Table 13.1 shows that the annual energy used for heating is broadly comparable to the electrical motive energy for refrigeration generation and the distribution systems. Table 13.2 shows significantly that for an air waste system the power used for distribution is almost as much as for water chilling and the air side is three times that of the water side.

Reverse cycle unitary system (fig 13.2)

This consists typically of a series of small electrically driven refrigeration terminal units with a constant supply of water at 80°F (27°C). The units act either as a heat pump, heating air by taking low grade heat from water, or as an air cooler by rejecting heat to the water. Ventilation air is supplied by a separate system usually at room temperature.

The central energy conversion plant supplies heat to the circulating water through a water-to-water heat exchanger, or rejects heat from the circulating water through a closed circuit cooling tower.

The system has the following energy conserving features

- Only the amount of air required for ventilation is distributed. This is a comparatively small amount and can be distributed at low velocity, thus the energy demand on supply and extract fans is minimised.

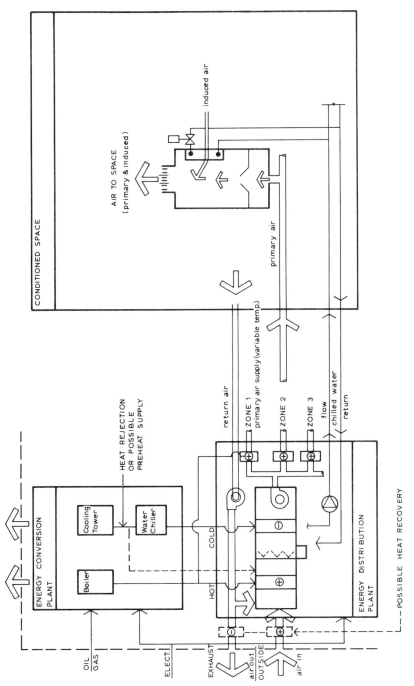

13.4 Two pipe induction system (non change-over)

- The system can virtually transfer heat and cold from one zone to another making it possible to take advantage of the diversity of loads.
- The circulating medium is water, not air. The temperature differential between the water and the spaces through which it passes is small, again minimising energy loss in distribution.
- The terminal units can be switched off when not required.

The disadvantage of the system is that the overall coefficient of performance of the refrigeration unit is of the order of 2.5 to 1 compared with a central chiller of 3.5 to 1. This in turn means that when in a heating mode about one third of the heat to the room is from an electrical source.

Four pipe fan coil system (fig 13.3)

This basic system consists of terminal units with a fan to recirculate air within the conditioned space. The units are served with hot and chilled water.

The chilled water and hot water temperatures to the building can be modulated according to seasonal temperatures. Control at the terminal unit is effected by either modulating the water flow rate to the coils in sequence or by dampering the recirculated room air to pass over one coil or other.

The energy conserving features of the system are:

- Minimum amount of air is distributed for ventilation.
- Water is the medium for distributing heating and cooling.
- Modulation of chilled water temperature upwards when possible permits the water chiller to work at an improved COP.
- Only the amount of heating or cooling required is taken from the water distribution systems at each terminal.
- The terminal units can be switched off when not required.

Manually operated speed control on the fans of the terminal units can reduce the heating or cooling output of the terminal unit but it does not produce a saving in electrical energy of the fan.

Two pipe non change-over induction system (fig 13.4)

In its simplest form this system has a constant quantity of primary air supplied at varying temperatures to each zone of the building together with chilled air at a constant temperature. Heating is provided by the primary air and individually controlled cooling by modulating the flow of chilled water through the coil of the terminal induction unit.

The amount of air that has to be distributed is some 2½ times the amount required for ventilation alone. The pressure at the nozzles of the terminal units required for induction depends on equipment selection. This is usually in the range of 1-in. WG to 2-in. WG. Selection of terminals has a direct effect on the fan total pressure. The system is not good on energy conserving grounds although steps can be taken in system design and equipment selection to minimise the inherent energy consuming aspects of the system. The energy disadvantages of the system compared with the previous systems are

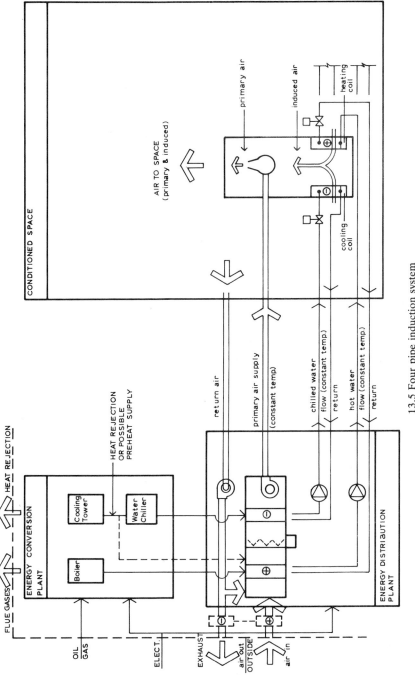

13.5 Four pipe induction system

- The medium for distributing heating is air.
- The quantity of air required means that it invariably has to be distributed at high velocity to minimise space requirements of ductwork. This leads to high friction losses in distribution.
- The requirement for high pressure at the nozzles of the terminal units also adds to the fan total pressure.
- The mixing losses in the terminal induction units. The induced air has to be cooled sufficiently to overcome any heating effect of the primary air to provide the correct supply air temperature to the spaces.
- Terminal units cannot be turned off when the building is only partially occupied.

Four pipe induction system (fig 13.5)
This is the most sophisticated and most expensive of the induction systems. A constant amount of primary air is supplied at room temperature to the terminal units throughout the occupied period. Constant amounts of hot water and chilled water are circulated at constant flow temperatures to the heating and cooling coils in each terminal unit.

Control of the temperature of air to the space is by means of sequential control of the water to the two coils or by damper control allowing the air to pass over either the heating or cooling coil, as with the four pipe fan coil system. In a sense this system is a combination of the two previous systems. In energy terms it is also a combination of the two systems. The central plant is virtually the same as for the two pipe induction system: the difference is that the heating medium is water rather than air.

The amount of air and the nozzle pressures required for induction are the same as for the two pipe system.

The energy advantages of the system are
- the medium for distributing most of the heating and cooling is water.
- There are no mixing losses at the terminals.

The disadvantages are as for the distribution losses of the two pipe induction system, and inability to turn off terminal units when not required.

All air system with terminal reheats (fig 13.6)
This system uses air to provide cooling and most of the heating to the conditioned space. The terminal reheats are purely used to give final temperature control to the spaces. All the air is supplied to and extracted from the individual spaces.

The amount of air required by this system can be from five to eight times the amount required by ventilation alone. Because of this and to give better control of the temperature conditions being supplied to each zone, it is usual to split the air distribution system into a series of air handling units located as near to the conditioned spaces as possible. In doing so the hot and cold water provided by the energy conversion plant has to be distributed through the building to the zone air handling plants.

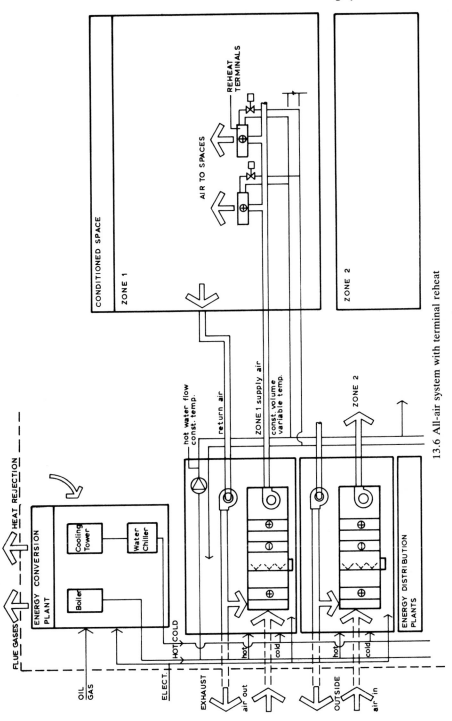

13.6 All-air system with terminal reheat

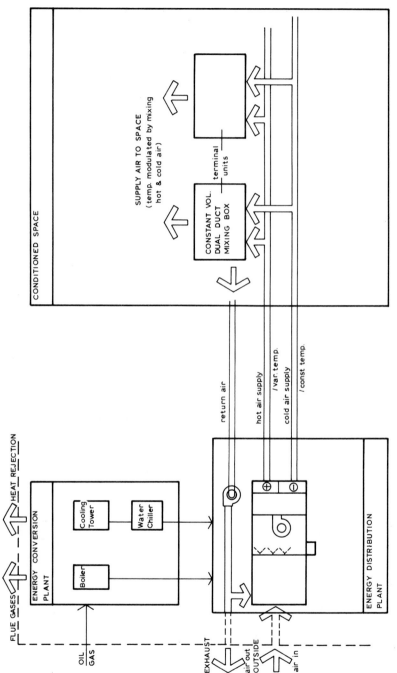

13.7 Dual duct system

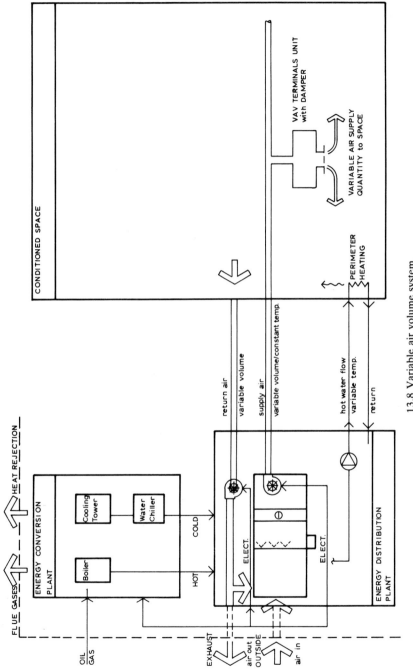

13.8 Variable air volume system

The system is inherently wasteful in energy terms and in fact is rarely used in large office developments. It can however be justified for small buildings or for special spaces.

The energy disadvantages of the system are:

- The air supplied to a zone must be at a temperature to give adequate cooling to the space with maximum cooling load. Spaces requiring any higher supply air temperature have to reheat this supply air to cancel the cooling.
- Air is the medium for both heating and cooling.
- The amount of air required means that high duct velocities are needed unless air handling units are located adjacent to conditioned spaces.

An energy advantage of the system is the ability to turn off the supply of air to a zone that is not occupied.

Dual duct system (fig 13.7)

In this all air system cooled air and heated air is distributed throughout the building and mixed at the terminal units.

The amount of air being distributed is again between five to eight times the amount required for ventilation alone and is distributed at high velocity.

The temperatures of the cooled and heated air are modulated to ensure minimum energy consumption. As much air as possible is recirculated.

The energy disadvantages of the system are:

- The losses in distribution.
- The mixing losses at terminals.
- It is not possible to turn off the supply to unoccupied spaces.

Variable air volume system (fig 13.8)

There are a number of ways in which the principle of this system is applied. Possibly the simplest is one in which the heat losses are offset by a perimeter heating system served by hot water, and cooling load is offset by a variable volume of cooled air.

The system overcomes many of the disadvantages of all air systems and properly applied can be comparable to the most energy conservative of the systems reviewed.

The energy advantages are:

- In the design of the air supply plant it is possible to take full advantage of the diversity of design heat gains and size down the central air handling plant. Although the design total air quantities can still be in the order of five to eight times that required for ventilation alone, it will be less than for the all air constant volume systems.
- Under partial load conditions the air quantity drops further. The average volume of air handled may be of the order of 50% of the design volume. As the energy consumed by the distributing fans is proportional to the cube of the volume of air handled the average energy consumed can be 30% to 40% of the design fan power.

• The heating medium is water, the temperature of which can be scheduled to match the heat losses.

CONCLUSION

The chapter has discussed the energy advantages and disadvantages of the seven types of air-conditioning systems in general terms. Whilst it is not possible to be specific about the order of merit of the systems in overall energy terms, the following order is suggested as reasonably representative.

1st, 2nd, 3rd Not in order –	Four pipe fan coil, reverse cycle unitary system, variable air volume systems
4th	Four pipe induction system
5th	Two pipe induction system
6th	Dual duct system
7th	Constant volume reheat system.

ACKNOWLEDGMENTS

I am most grateful for the guidance given me in the preparation of this chapter by Peter Jones of Haden Young Ltd, and to the Renton Howard Wood Levin Partnership for assistance in the preparation of illustrations.

REFERENCES

Energy Conservation: a study of energy consumption in buildings and possible means of saving energy in housing, BRE Working Party Report (Sources and Uses of Primary Energy in UK)
Electricity Council, *Energy Consumption of Air Conditioned Buildings,* October, 1973
W. P. Jones, Designing Air-Conditioned Buildings to Minimise Energy Use, (Paper to Nottingham Conference, 1974, 'Integrated Environment in Building Design')
IHVE, *Guide Book B,* 1970
N. O. Millbank et al, Investigation of Maintenance and Energy Costs for Services in Office Buildings, *JIHVE,* October, 1971
ASHRAE Handbook, 1976 Systems
J. C. Knight, L. G. Hadley, *Building Services: An Energy Demand Review*

DISCUSSION

D. Thornley (J. Roger Preston & Partners)   In 1964 in conjunction with Messrs Swain and Wensley* I undertook an exercise similar to that

*C. P. Swain, D. L. Thornley, R. Wensley. The Choice of Air-Conditioning Systems, *JIHVE,* vol 32, April 1964, p 307.

undertaken by Mr Goldfinger. In those days we were not so much concerned with energy conservation as monetary conservation and so our examination was based on the interest of owning and operating cost. Variable air volume systems were not then available in the UK but with that exception we covered most of the systems and it will be very interesting to go back and compare Peter Goldfinger's order of importance based on energy conservation with our order based on owning and operating cost. The order is not in fact very different which leads to the conclusion that saving energy saves money or in reverse, if you save money you will save energy. Of course there are exceptions to this, if we have big plant investments, but I think perhaps there is a general message to be learned.

D. Oughton (Oscar Faber & Partners)   Is the ranking of air-conditioning systems based on the consumption of site or national energy?

J. P. G. Goldfinger (Dale & Ewbank)   The ranking is based on site energy. That is the energy that is interesting to the building owner.

14 Heat pumps: the future

EDWARD JAMES PERRY

INTRODUCTION

It is the duty of all engineers and plant owners to carry out a techno-economic appraisal of any energy using system during the conceptual design stage of a project and before any contract is placed. To achieve this objective the capital and running costs of the whole complex must be examined for all options that would fulfil the project's function. This necessitates taking into account the availability and cost of the various forms of energy that could be used by the plant and be available during the life of the installation.

Since the world's fossil fuel reserves are limited which in turn will escalate energy costs at a greater rate than the average inflation rates, the fuel and power costs of operating plant will become more important as the fuel reserves decline. Therefore, any energy-using system should be such that it conserves energy and is economical in total cost. The heat pump is one such device that can, if correctly designed and applied, achieve both economies in fuel and total costs.

The degree to which the heat pump is developed and used over the years depends upon which energy route we pursue as a nation. We could, in the future, rely mainly on nuclear power for our energy and therefore the heat pump would then become a very important energy conversion component. On the other hand, we could utilise the nation's existing natural gas network for distributing SNG when the heat pump would be used to a lesser degree. Irrespective of these two scenarios the heat pump, as developed at the present time, achieves considerable savings in energy conservation and, in many cases, operating costs.

The heat pump principle is applied to many industrial processes and to some building services systems. In both areas, development of equipment is required and in the case of building service applications, a better understanding of the instantaneous interrelationship of heating and cooling loads and a departure from catalogue engineering is required.

TYPES OF HEAT PUMPS

There are three basic forms of heat pumps, these being the better understood 'closed cycle' system, fig 14.1a, the 'open cycle', fig 14.1b and the absorption system, fig 14.1c. The open and closed cycle systems will only be considered in this paper as the energy and economic benefits obtainable from the absorption system are limited.

Open and closed cycle heat pumps have many suitable applications in the process industries and those industries serviced by the building services engineers. The following examples may help identify areas where significant improvements in energy utilisation and cost benefits can be achieved.

- product drying such as for paper and construction materials
- product drying of grain and agricultural products
- distillation of liquids
- environmental heating and cooling of buildings

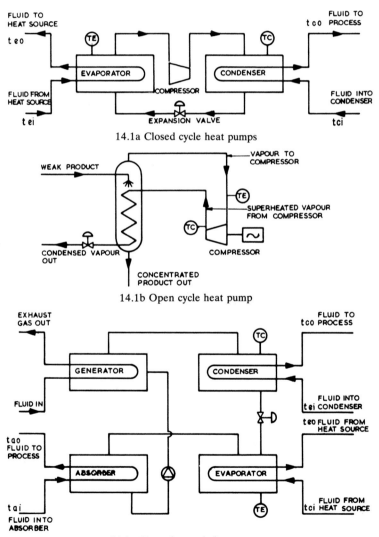

14.1a Closed cycle heat pumps

14.1b Open cycle heat pump

14.1c Aborption cycle heat pumps

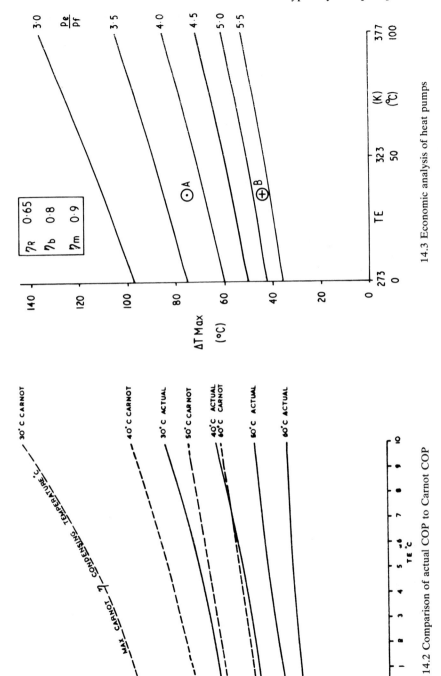

14.3 Economic analysis of heat pumps

14.2 Comparison of actual COP to Carnot COP

- utilisation of waste heat from refrigeration plants for process or environmental heating
- utilisation of waste heat from power stations for environmental heating
- utilisation of heat extracted from ice rinks to heat swimming pools.

DESIGN CRITERIA AFFECTING PERFORMANCE

The actual coefficient of performance of a heat pump is defined by:

$$COP_{act} = \frac{\text{condenser heat}}{\text{absorbed compressor power}}$$

While the theoretical maximum COP is defined by:

$$COP_m = \frac{T_c}{(T_c - T_e)}$$

where T_c is the absolute condensing temperature and T_e the absolute evaporating temperature.

The relationship between COP_m and COP_{act} can be expressed as the Carnot efficiency η_r and for the range of actual COP to the maximum COP given in fig 14.2, averages at 0.65 for a single stage R12 installation. If a two stage compressor was used in place of the single stage unit for the temperature range given in fig 14.2, then the Carnot efficiency would increase to approximately 0.75.

It can be seen from fig 14.2 that the COP is sensitive to the temperature stretch over which the heat pump operates and to the number of compression stages incorporated in the system. This difference in temperature, $T_c - T_e$ should be a minimum for a maximum COP which implies that the terminal temperature differences $T_e - t_{eo}$ and $T_c - t_{eo}$ of the evaporator and condenser should be as small as practicable. Therefore, on large installations heat exchange surfaces and compressor systems should be optimised relative to energy costs and plant utilisation.

The coefficients of heat transfer for liquids are greater than those of gases, which generally means that for the same capital expenditure on the heat exchanger small approach temperatures can be achieved. The increased density and specific heat characteristics of liquids to those of gases also have a similar effect.

Future heat pumps must take more advantage of any liquid or condensing source of low grade heat in preference to, say, air so as to maximise the COP with a minimum capital expenditure.

ENERGY COST COMPARISONS

If a heat pump is not part of a refrigeration system then the energy costs of operating the plant must be less than say those of a boiler plant. Since the economics of a heat pump are dependent upon the operating temperatures and temperature difference of T_e and T_c an equation has been derived (1) for the maximum economic temperature rise from the heat source to the heating fluid as,

$$\Delta T_{max} = \eta_r \, T_e \Big/ \left(\frac{P_c \eta_b}{P_f \, \eta_m} \right) - 10 \qquad (1)$$

where ΔT_{max} = $T_c - T_e$
η_r = Carnot efficiency
η_b = boiler efficiency
η_m = motor efficiency including power factor
P_c = electrical power costs
P_f = fuel costs
T_c = condensing temperature °K
T_e = evaporating temperature °K

This equation assumes that the temperature approach of the condenser is 5°C to their respective fluids. This equation has been evaluated for a range of parameters and the results are plotted in fig 14.3. It can be seen from this figure that the maximum economic temperature rise across a heat pump increases as the ratio of the electricity price to the boiler fuel price falls. The use of the figure may be demonstrated by a simple example if steam is required to be raised by a heat pump at 110°C by using cooling tower water say at 35°C then the required temperature rise is 75°C this point is marked A on fig 14.3. In order for the scheme to be viable, the ratio of electricity price to the boiler fuel price would have to be less than 3.8. This ratio based on electricity charges at 2p/kWh and gas at 11p/therm currently stands at 5.3. It is therefore clear that raising steam from low grade waste heat is uneconomic at the present time but this ratio will change in the future. The maximum economic temperature based on the energy costs given in the above example is marked B in fig 14.3 and equates to a temperature of 77°C, a temperature rise of 42°C above the cooling tower water temperature of 35°C.

RELATIONSHIP OF HEATING AND COOLING LOADS

Condenser heat reclaim system of air-conditioning plants can only supply heat in direct ratio to the heat extracted from the building. In many cases the heat requirements may be greatly in excess of the refrigeration load and therefore additional heat must be provided from a fuel fired source, or alternatively by using the waste heat to provide the load for the heat pump system. Fig 14.4 shows diagramatically how a standard refrigeration plant can be economically converted into a heat pump system with a water source. The limitations are that the compressor is either reciprocating or screw and has a motor capable of operating with the increased condensing load. Care

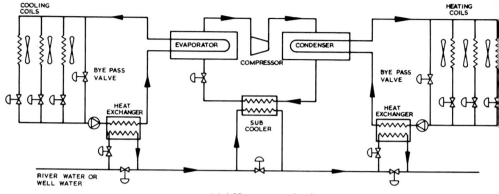

14.4 Heat pump circuit

should be taken with centrifugal machines to ensure that the operating parameters do not place the compressor in the surge regime.

This heat pump system has the advantage of maintaining a closed circuit for the heating and cooling circuits while the river, sea, well or waste process water circuit heat exchangers can be of the plate type. These units have the advantage that they can be easily cleaned without disturbing the pipework. With heat pumps, it is highly desirable to take advantage of the cooler river or well water to make the thermodynamic cycle more efficient by incorporating liquid refrigerant sub-coolers. This minimises the power requirements of the compressor for a given cooling and heating duty.

The use of computer simulation of building loads and energy systems enables instantaneous interrelationship of both the cooling and heating requirements to be established. This enables the annual operating costs of various energy systems to be financially assessed during the design stage of a project. With this added confidence in predicting the behaviour and running costs of such systems, more installations will be installed in the future.

GROWTH OF HEAT PUMP APPLICATIONS

The use of heat pumps in the UK has been limited due to the availability of cheap energy and to quote the July 1977 issue of ASHRAE, 'the inability of UK engineers to obtain the instantaneous relationship of heating and cooling demands of air conditioning loads'. In addition to these reasons very often some building services engineers rely too heavily upon catalogue data and do not give adequate thought to the operating cycles and costs of such systems. The individual chilling units are usually well engineered but when more than one such unit is incorporated into a complete system the interaction of load changes and external control loops in the air-handling plants can and do cause operating difficulties and waste of power.

To illustrate the growth of heat pump applications in the continent, Sulzers heat pumps reference list has been analysed for the last 40 years. Fig

14.5 shows the installed load against time for various types of heat pumps and applications. It is interesting to note that the general shape of the curve is similar to the curve published by ASHRAE for unitary heat pumps. This curve is shown in fig 14.6.

It should be noted that by far the biggest growth rate of heat pumps on the continent is the system utilising water as the source of heat, particularly well water, as this has the added advantage of maintaining a relative constant temperature throughout the year of about 9 to 10°C. This can be compared to the annual fluctuations of the dry bulb temperature of the air, say, from −3 to 30°C. This would result in a lower COP due to the reduced evaporating temperature in the winter when the maximum load is required.

Other sources of heat include the earth, where evaporator coils are buried at sufficient depth to be unaffected by the surface temperatures. Alternatively, solar heated panels can be used as the heat source. The solar panel can

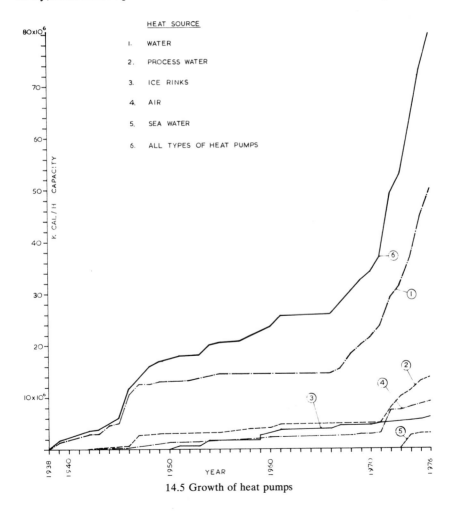

HEAT SOURCE

1. WATER
2. PROCESS WATER
3. ICE RINKS
4. AIR
5. SEA WATER
6. ALL TYPES OF HEAT PUMPS

14.5 Growth of heat pumps

heat water for storage and the heat pump can raise the stored water temperature to a usable level. Alternatively, the evaporator can be incorporated into the solar panel. This can then be used as either a heat pump or a heat pipe depending on whether the solar panel temperature is greater than the required hot water temperature. In this latter mode of operation the compressor would be by-passed.

Heavy vapour laden air such as exhausts from swimming pools are suitable sources of heat as the heat transfer characteristic and the latent heat component of the air contribute to a higher COP.

ECONOMIC ANALYSIS OF A HEAT PUMP SYSTEM

Since the future of heat pumps depends upon their correct application and economic viability a summarised analysis of a study for such a system is given here to highlight important factors such as utilisation, ratio of energy requirements and tariffs.

The study used as the example was for investigating the economics and the savings in energy of heat pump systems compared to conventional methods of supplying the heating requirements of a school building built with good insulation standards. Eight options were investigated, but only four will be explained. For options 1, 2 and 3 the heat source was river water.

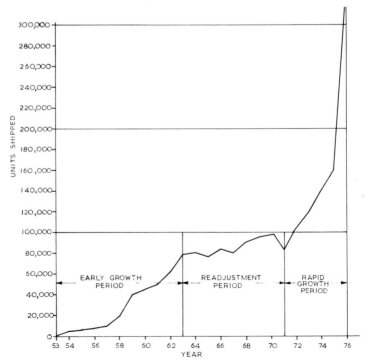

14.6 Industry shipments of unitary heat pumps

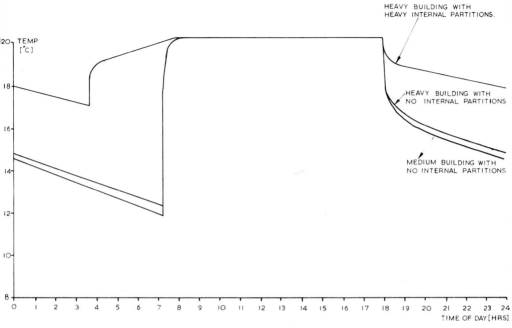

14.7 Typical curve showing effect of building structure on daily temperature profiles

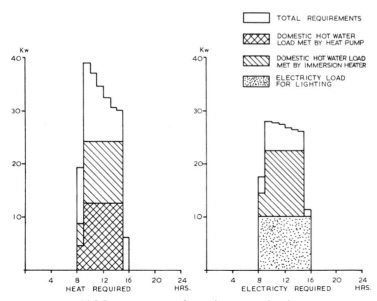

14.8 January average day optimum start heating

Option 1   Lighting electric, space and water heating, off-peak electric water to water heat pump with storage. Off-peak immersion for water temperature boost, cooking electric.

Option 2   Lighting electric, space and water heating, off-peak electric water to water heat pump with storage. Off peak immersion for water temperature boost, gas cooking.

Option 3   Lighting electric, space and water heating and cooking by gas.

Option 4   Lighting electric, space heating electric air to air heat pumps, water heating and cooking by gas.

The method of analysis was to:

1. Determine the daily thermal and energy demand profiles on an hourly basis for the year taking into account fixed time starts for heating.
2. Determine domestic hot water requirement profiles allowing for storage.
3. Determine cooking and lighting loads.
4. Determine heat pumps and boiler plant energy requirements allowing for plant efficiency and taking into account hourly temperature changes of the heat source.
5. Determine annual operating costs of each system.
6. Calculate net present values of cash requirements to finance capital sums plus operating costs discounted at 10% over 20 years.
7. Determine the prime energy requirements for space heating and of the complete installation.

A typical curve showing the effect of the building structure mass on the daily internal temperature is shown in fig 14.7 for intermittent heating. The daily heat demand and power requirement profiles for the school for an average January day are shown in fig 14.8. Simulating the hourly variations of heat demands over a year of school activity taking into account the weekends and holidays, a load duration curve can be generated, see fig 14.9. It can be seen that the maximum heating load occurs for a very short period with 50% of the load occurring for only approximately 10% of the year.

A schematic diagram of the water to water heat pump is given in fig 14.10. A Baudelot cooler was selected so that the heat exchanger surface could be cleaned without dismantling components.

A summary of the results are given in tables 14.1, 14.2 and 14.3. Table 14.1 quantifies the fuel costs for the different options. It can be seen that the lighting costs are more in all cases than the space heating costs which in itself is less than the costs of the hot water duty. It should be noted that the least heating fuel cost is the water to water heat pump options 1 and 2. Tariffs used are quoted in reference (5).

When the electric cooking load is included on option 1 the water to water heat pumps become more expensive. Table 14.2 shows the various capital cost elements of the four options. It can be seen that the water to water heat pump is more expensive in capital costs, while it is the least expensive in fuel costs. Table 14.3 shows the analysis of the capital and running costs showing that option 3 is by far the cheaper in total cost.

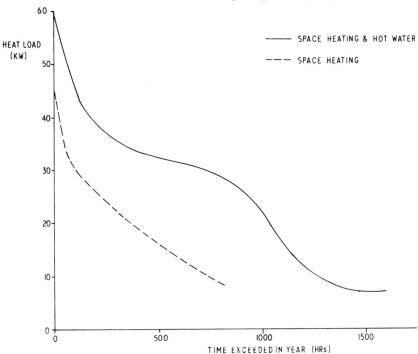

14.9 Load duration curve

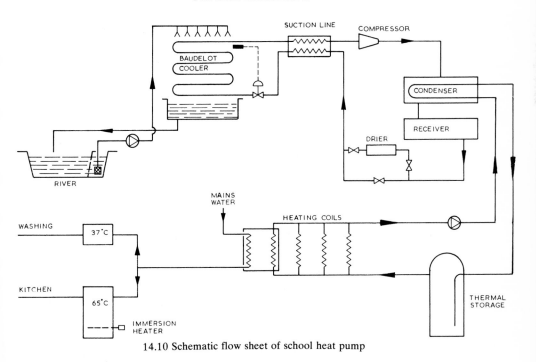

14.10 Schematic flow sheet of school heat pump

Table 14.1   Fuel costs options 1–4

	Option 1 Water to water heat pump + immersion off-peak water heating. Electric cooking	Option 2 Water to water heat pump + immersion off-peak water heating. Gas cooking	Option 3 Gas boiler. Gas cooking	Option 4 Air to air heat pump+gas water boiler + gas cooking
Lighting	347	347	347	347
Space heating	42	42		268
Space heating gas			128	
Water heating	37 }210	37 }210		
Electricity	173 }	173 }		
Gas	–	–	226	226
Total heating	252	252	354	494
Total	599	599	701	841
Cooking electric	569			
Cooking gas		229	229	229
Total fuel cost £/A	1168	828	930	1070

This economic analysis indicates the importance of investigating all factors affecting the total operation of a building. It also shows that one cannot make policy decisions by reviewing the operating and capital costs of only one aspect of the building. The study also shows how important the utilisation of plant is in making more complex schemes economical.

HEAT PUMP TEMPERATURE RANGES

To deliver heat equivalent to the heat input at the power station level the COP of a heat pump would have to be approximately 3.7, see Sankey diagram, fig 14.11a. If waste heat from a heat source together with waste

Table 14.2   Capital costs — options 1–4

	Option 1 Water to water heat pump+immersion off peak water heating. Electric cooking	Option 2 Water to water heat pump+immersion off peak water heating. Electric cooking	Option 3 Gas boiler. Gas cooking	Option 4 Air to air heat pump+gas water boiler + gas cooking
	£	£	£	£
Mechanical equipment	20,000	20,000	4800	9000
Heat pump ancillaries	4800	4800	—	—
Ductwork	—	—	—	5000
System installation	4000	4000	3000	3300
Gas connection charge	—	6400	6400	6400
Electrical connection charge	1000	1000	1000	1000
Total	29,800	36,200	15,200	24,700

Table 14.3 **Net present costs and primary energy requirements options — 1–4**

Scheme	Option 1 Water to water heat pump+ immerson off peak water heating+electric cooking	Option 2 Water to water heat pump+ immersion off peak water heating+gas cooking	Option 3 Gas boiler. Gas cooking	Option 4 Air to air heat pump+gas water boiler+ gas cooking
Fuel costs	1168	828	930	1070
Maintenance	600	600	300	500
Total running	1768	1428	1230	1570
Running NP cost	15,050	12,156	10,470	13,365
Capital cost	29,800	36,200	15,200	24,700
Total NP cost	44,850	48,356	25,670	38,065
Rank	3	4	1	2
Prime Energy requirements MWh	235	200	181	187
Total Space heating	15	15	31	32

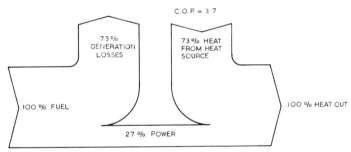

14.11a Electric powered heat pumps

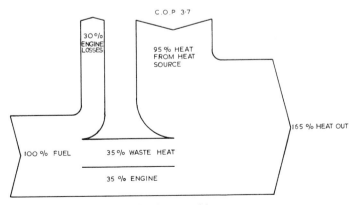

14.11b Diesel powered heat pumps

diesel heat, 165% heat output could be obtained for a 100% fuel input as shown in fig 14.11b.

Figure 14.12 has been prepared giving maximum Carnot and maximum theoretical temperature limits that heat pumps can raise the heat source temperature for varying COP's. For example, if the temperature of a heat source was 10°C the maximum temperature that can be delivered by a heat pump against this cycle would be 70°C assuming a 5°C approach temperature for each heat exchanger. If the heat source was at a temperature of 100°C then an exit temperature of 180°C could be expected.

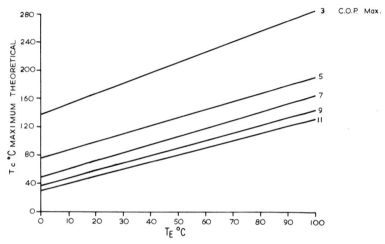

14.12a Carnot maximum condensing temperatures for varying evaporating temperatures and COP

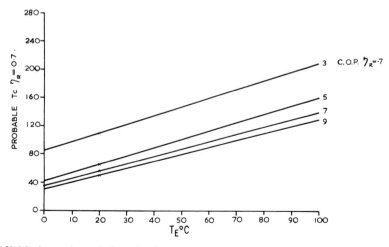

14.12b Maximum theoretical condensing temperatures for varying evaporating temperatures and COP

Units are already in production capable of delivering heat at a temperature of 105°C with a water source temperature of 64°C having a COP of 3.7. This would indicate that the terminal temperature differences are in the order of 15°C.

Studies are being carried out in the USA by Neil and Jensen to recover geothermal heat by means of a heat pump, also work is being carried out to raise the power station cooling water temperature by means of heat pumps to more suitable temperatures for distribution without losing power station efficiency.

It would appear that benefits could be achieved in air-conditioning plants by producing ice on ice banks or as a slush and use the ice to produce the chilled water. This method would help produce process heat when cooling was not required and cooling when heating was not demanded. This method can level out peak loads and is used in process industries. Great care must be taken to ensure that the latent thermal storage capacity and surface is compatible with the maximum thermal peaks and their duration.

FUTURE DEVELOPMENTS

Future heat pump developments will concentrate on utilising better sources of heat. These will be well, river and process water systems. This policy may necessitate using purpose made heat pump systems and, to increase the COP, multi-stage compressors will be incorporated into the design. Heat storage using solar energy systems is another source for heat pump development, particularly in countries that have a bigger incidence of sunshine than the UK.

With the increased insulation standards in private dwellings, and building construction methods being such as to exclude draughts, the development of small domestic heat pump systems is a possibility. The heat source could be the ground underneath the house, a heat exchanger being incorporated into the foundations during the construction of the dwelling.

Compressors and refrigerants will be developed to suit the higher operating temperatures. Heat pumps driven by diesel or gas turbines will be developed in the future using the waste heat from the engine to raise the temperature of the heat source. In such systems greater temperature rises will be obtainable.

Further development for very large installations will be to make use of mixed refrigerant systems that are matched to the temperature gradient of the heat source that is to be cooled. Above all, the dynamics of all heat pump systems should be analysed so as to ensure economic success in operation.

In drying applications such as for paper, foodstuffs, textiles etc, the use of direct and indirect heat pumps will become areas where operating costs can be drastically reduced. Building service engineers are becoming more involved in applications, and should be aware of such developments.

REFERENCES

1 E. J. Perry and M. V. Casey, Energy and Refrigeration Systems–Design Maintenance and Operation, *Proceedings of the Institution of Refrigeration,* London, 1976–7.
2 *ASHRAE J,* July, 1977.
3 J. Lawton, J. E. T. MacLaren and D. C. Freshwater, The Watt Committee on Energy: Heat Pumps in Industrial Processes, Institution of Mechanical Engineers, June, 1977.
4 D. T. Neill and W. P. Jenson, *Geothermal Powered Heat Pumps to Produce Process Heat,* 1976.
5 South Western Gas Single Part Tariff.
  South Western Electricity Board Off Peak, maximum demand and cooking tariff.

ACKNOWLEDGEMENT

The author thanks W. S. Atkins & Partners for permission to publish.

DISCUSSION

D. Thornley (J. Roger Preston & Partners)   Mr Perry seemed to be selling heat pumps short by his choice of figures for current costs of electricity and fossil energy. Recent and projected figures suggest that the price ratio of electricity to boiler fuel is close to 3.8 already. Our practice is engaged in a large project in Hong Kong where we are absolutely convinced there is a case for using the refrigeration plant with sea water which is conveniently never below 15°F with plate heat exchangers. As fuel prices increase the case would be even stronger.

I would finally like to quote a comment about the importance of heat pumps research, 'Certainly the present fuel situation in this country is sufficiently grave to warrant intensified research on a national scale comparable with that which has been entered into in the field of atomic energy and gas turbine development'. It was said to the IHVE in their summer meeting, 1948.

E. J. Perry (W. S. Atkins & Partners)   The energy costs and fuel costs used were taken from industrial tariffs but it must be borne in mind that it is the ratio of electricity cost of fuel costs that is the important factor and I would doubt if the ratio would alter greatly even if domestic tariffs were used.

W. F. Stanton (How Group Southern Ltd)   I would like to ask Mr Perry if, for the example of the heat pump that he described, he would give an indication of the temperature conditions on the Baudelot cooler and the condenser side. Also whether the system met the minimum COP of 3.7 that he indicated was required.

E. J. Perry The evaporating temperature was of the order of 2°C, the condensing temperature 46–49°C with the water being circulated to the heating units at 43°C. Statistical data showed the river water never froze and consequently we took advantage of a large water flow and only 1–2°F being taken out of the river water.

M. B. Ullah (Robert Gordons Institute of Technology) From the analysis it appears that diesel powered heat pumps may be competitive in terms of working cost, but the fact remains these are not used. Could Mr Perry indicate some reasons why?

E. J. Perry People have doubts upon the reliability of diesel engines therefore they tend not to install them on heat pump systems. To obtain independent reliability figures other than the manufacturers' the Diesel Users Association publish recorded data of down time, periods between overhaul and failure rates. From this information you can establish the reliability factors which can relate to your actual plant loading and can, therefore, make judgment on the degree of standby required. This data is also published relative to gas turbine drives.

J. J. Wilson (Lloyd's Register of Shipping, Institute of Refrigeration) Diesel engines do give reliable service, generally, provided they are maintained and serviced correctly and are not run continuously at maximum output.

This has been confirmed by Lloyd's Register of Shipping's compiled records from surveyors' detailed reports of failures and defects, the most comprehensive of their kind in the world.

At a number of Cold Stores classed with Lloyd's Register, at places like Ghana, Nigeria and several countries in the Middle East, continuity of electrical power from the main supply cannot be relied upon and in these cases we recommend diesel generators be available.

A number of generator sets are obviously better than one and if the generator sets are all the same model, one or more could be cannibalised for spares if the necessity arose.

Obtaining spares in some countries could be a problem and we advise supply of these be kept available.

In the early days, some diesel engines were considered to be temperamental, but most provide a commendable degree of reliability provided they receive the correct treatment.

Index

284